Mind and Brain

Mind and Brain

. . .

READINGS FROM
SCIENTIFIC AMERICAN MAGAZINE

W. H. FREEMAN AND COMPANY
New York

Library of Congress Cataloging-in-Publication Data

Mind and brain : readings from Scientific American magazine.
p. cm.
"The eleven chapters and the epilogue in this book originally appeared as articles and the closing essay in the Sept. 1992 issue of Scientific American".—T.p. verso.
Includes bibliographical references and index.
ISBN 0-7167-2376-X
1. Cognitive neuroscience. 2. Brain.
3. Neurophysiology.
I. Scientific American.
QP360.5.M55 1993
152—dc20 92-34725
CIP

The eleven chapters and the epilogue in this book originally appeared as articles and the closing essay in the September 1992 issue of SCIENTIFIC AMERICAN.

Printed in the United States of America

1 2 3 4 5 6 7 8 9 0 RRD 9 9 8 7 6 5 4 3 2

CONTENTS

Foreword

What is the relation between the mind and the brain? René Descartes believed the brain was an extracorporeal entity that gained access to the individual and the surrounding world through the pineal gland. For many centuries, other students of the question shared Descartes's belief in the existence of an independent, conscious entity. How, after all, could something so exquisitely abstract, so unitary as one's awareness of being aware arise from the bilobed, ridged, and folded organ encased in the human skull? The work of Santiago Ramón y Cajal and the generations of neurobiologists who have followed him began to suggest some answers to that rhetorically framed question.

The growing body of anatomy and electrophysiology that they created complemented animal studies and the clinical observation of individuals whose injured brains offered sometimes bizarre, tragic evidence of a relation between structure and function. Today magnetic resonance imaging, positron emission tomography, and other analytical techniques have extended the reach of the traditional methods of inquiry. The effort has begun to uncover the physiological foundations of the mind. Such attributes of the mind as the capacity to see, use language, learn, and remember can be understood in terms of physiological structure and function at the cellular and subcellular levels. Even the electrophysiological correlates of consciousness seem to be within experimental reach.

The emerging body of knowledge offers much that fascinates, illuminates, and even disturbs. Although the brain's architecture takes shape in the fetus, it can assume its final form only through such stimuli as touch, speech, and image. The world we see is literally the creation of our brain: we are most definitely not cameras. A set of rules by which the connections between neurons are strengthened underlies memory. A growing body of research indicates that the differences in perceptual style and capacity that psychology identifies in males and females have their roots in the physiology of the brain.

All this excellent science raises still more philosophical questions. Among them is whether the mind can ever finally know itself or whether it will, like the hare in Xeno's paradox, stay a fraction of a jump ahead of its pursuer. The study of mind and brain also suggests that a certain existential ambiguity colors existence. A common cerebral architecture unites all individuals. Yet because each person's experience is unique, so is each brain. Therefore, each individual lives in a world no one else can inhabit.

Mind and Brain

Mind and Brain

Introducing an overview of the emerging biological foundations of consciousness, memory and other attributes of mind—the most profound of all research efforts.

• • •

Gerald D. Fischbach

Ruth gave me a piece of her mind this morning. I am grateful, of course, but I don't know where to put it or, for that matter, what it is. I suppose that the imperatives belong in the limbic system and the geographic information in the hippocampus, but I am not sure. My problem also troubled René Descartes. Three centuries ago he described the mind as an extracorporeal entity that was expressed through the pineal gland. Descartes was wrong about the pineal, but the debate he stimulated regarding the relation between mind and brain rages on. How does the nonmaterial mind influence the brain, and vice versa?

In addressing this issue, Descartes was at a disadvantage. He did not realize the human brain was the most complex structure in the known universe, complex enough to coordinate the fingers of a concert pianist or to create a three-dimensional landscape from light that falls on a two-dimensional retina. He did not know that the machinery of the brain is constructed and maintained jointly by genes and by experience. And he certainly did not know that the current version is the result of millions of years of evolution. It is difficult to understand the brain because, unlike a computer, it was not built with specific purposes or principles of design in mind. Natural selection, the engine of evolution, is responsible.

If Descartes had known these things, he might have wondered, along with modern neurobiologists, whether the brain is complex enough to account for the mystery of human imagination, of memory and mood. Philosophical inquiry must be supplemented by experiments that now are among the most urgent, challenging and exciting in all of science. Our survival and probably the survival of this planet depend on a more complete understanding of the human mind. If we agree to think of the mind as a collection of mental processes rather than as a substance or spirit, it becomes easier to get on with the necessary empirical studies. In this context the adjective is less provocative than the noun.

The authors of the chapters in this book and their colleagues have been pressing the search for the neural basis of mental phenomena. They assume that mental events can be correlated with patterns of nerve impulses in the brain. To appreciate the meaning of this assumption fully, one must consider how nerve cells, or neurons, work; how they communicate with one another; how they are organized into local or distributed networks, and how the connections between neurons change with experience. It is also important to define clearly the mental phenomena that need to be explained. Remarkable advances have been made at each level of analysis. Intriguing correlations have in fact begun to emerge between mental attributes and the patterns of nerve impulses that flare and fade in time and space, somewhere inside the brain.

The most striking features of the human brain are the large, seemingly symmetric cerebral hemispheres that sit astride the central core, which extends down to the spinal cord. The corrugated hemispheres are covered by a cell-rich, laminated cortex two millimeters in thickness. The cerebral cortex can be subdivided by morphological and functional criteria into numerous sensory receiving areas, motor-control areas and less well-defined areas in which associative events take place (see boxed figure "The Brain: Organ of the Mind"). Many observers assume that here, in the interface between input and output, the grand syntheses of mental life must occur.

It may not be that simple. Mind is often equated with consciousness, a subjective sense of self-awareness. A vigilant inner core that does the sensing and moving is a powerful metaphor, but there is no a priori reason to assign a particular locus to consciousness or even to assume that such global awareness exists as a physiologically unified entity. Moreover, there is more to mind than consciousness or the cerebral cortex. Urges, moods, desires and subconscious forms of learning are mental phenomena in the broad view. We are not zombies. Affect depends on the function of neurons in the same manner as does conscious thought (see Figure 1.1).

And so we return to the organ itself. The brain immediately confronts us with its great complexity. The human brain weighs only three to four pounds but contains about 100 billion neurons. Although that extraordinary number is of the same order of magnitude as the number of stars in the Milky Way, it cannot account for the complexity of the brain. The liver probably contains 100 million cells, but 1,000 livers do not add up to a rich inner life.

Part of the complexity lies in the diversity of nerve cells, which Santiago Ramón y Cajal, the father of modern brain science, described as "the mysterious butterflies of the soul, the beating of whose wings may some day—who knows?—clarify the secret of mental life." Cajal began his monumental studies of adult and embryonic neurons about 100 years ago, when he came across Camillo Golgi's method of staining neurons with silver salts (see Figure 1.2). The great advantage of this technique, which led Cajal to his neuron doctrine, is that silver impregnates some cells in their entirety but leaves the majority untouched. Individuals thus emerged from the forest. Seeing them, Cajal realized immediately that the brain was made up of discrete units rather than a continuous net. He described neurons as polarized cells that receive signals on highly branched extensions of their bodies, called dendrites, and send the information along unbranched extensions, called axons. The Golgi stain revealed a great variety of cell-body shapes, dendritic arbors and axon lengths (see Figure 1.3). Cajal discerned a basic distinction between cells having short axons that communicate with neighbors and cells having long axons that project to other regions.

Shape is not the only source of variation among neurons. Diversity is even greater if molecular differences are considered. Whereas all cells contain the same set of genes, individual cells express, or activate, only a small subset. In the brain, selective gene expression has been found within such seemingly homogeneous populations as the amacrine cells in the retina, the Purkinje cells in the cerebellum and the motor neurons in the spinal cord. Beyond the structural and molecular differences, even more refined distinctions among neurons can be made if their inputs and projections are taken into account. Is it possible that each neuron is unique? This is certainly not the case in all but the most trivial circumstances. Yet the fact that the brain is not made up of interchangeable parts cannot be ignored.

In the face of this astounding diversity, it is a relief to learn that simplifications can be made. Several years ago Vernon B. Mountcastle, working on the somatosensory cortex, and David H. Hubel and Torsten N. Wiesel, working on the visual cortex, produced an important insight. They observed that neurons of similar function are grouped together in columns or slabs that extend through the thickness of the cortex. A typical module in the visual cortex whose component cells respond to a line of a particular orientation measures approximately one tenth of a millimeter across. The module could include more than 100,000 cells, the great majority of which participate in local circuits devoted to a particular function.

Another simplification is that all neurons conduct information in much the same way (see boxed figure "How Neurons Communicate"). Information travels along axons in the form of brief electrical

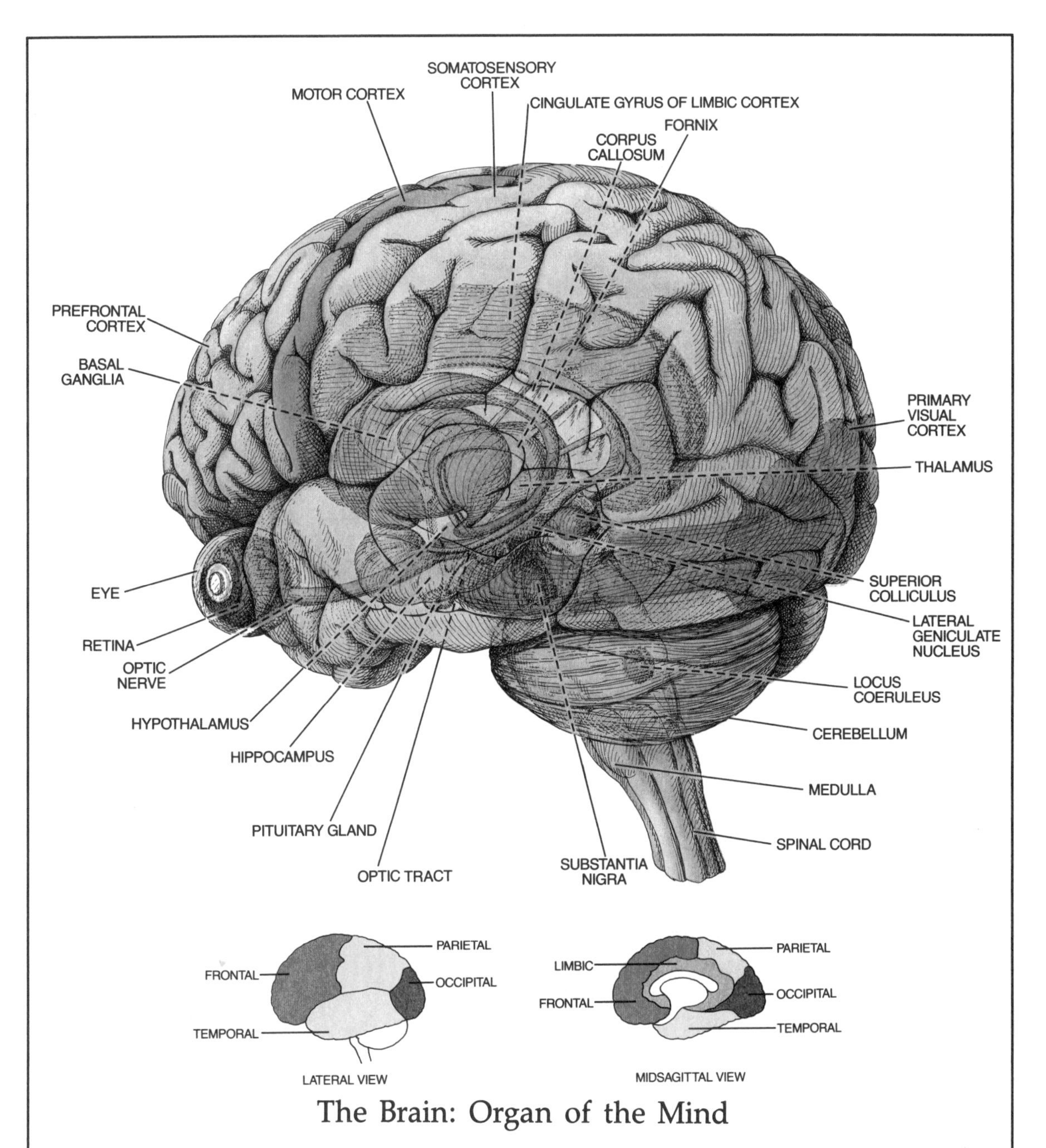

The Brain: Organ of the Mind

At a very gross level, the brain is bilaterally symmetric, its left and right hemispheres connected by the corpus callosum and other axonal bridges. Its base consists of structures such as the medulla and the cerebellum. Within lies the limbic system (*blue*), a collection of structures involved in emotional behavior, long-term memory and other functions. The most evolutionarily ancient part of the cortex is part of the limbic system. The larger, younger neocortex is divided into frontal, temporal, parietal and occipital lobes. Most thought and perception take place as nerve impulses, called action potentials, which move across and through the cortex. Some brain regions with specialized functions are the motor cortex (*pink*), the somatosensory cortex (*yellow*) and the visual pathway (*purple*).

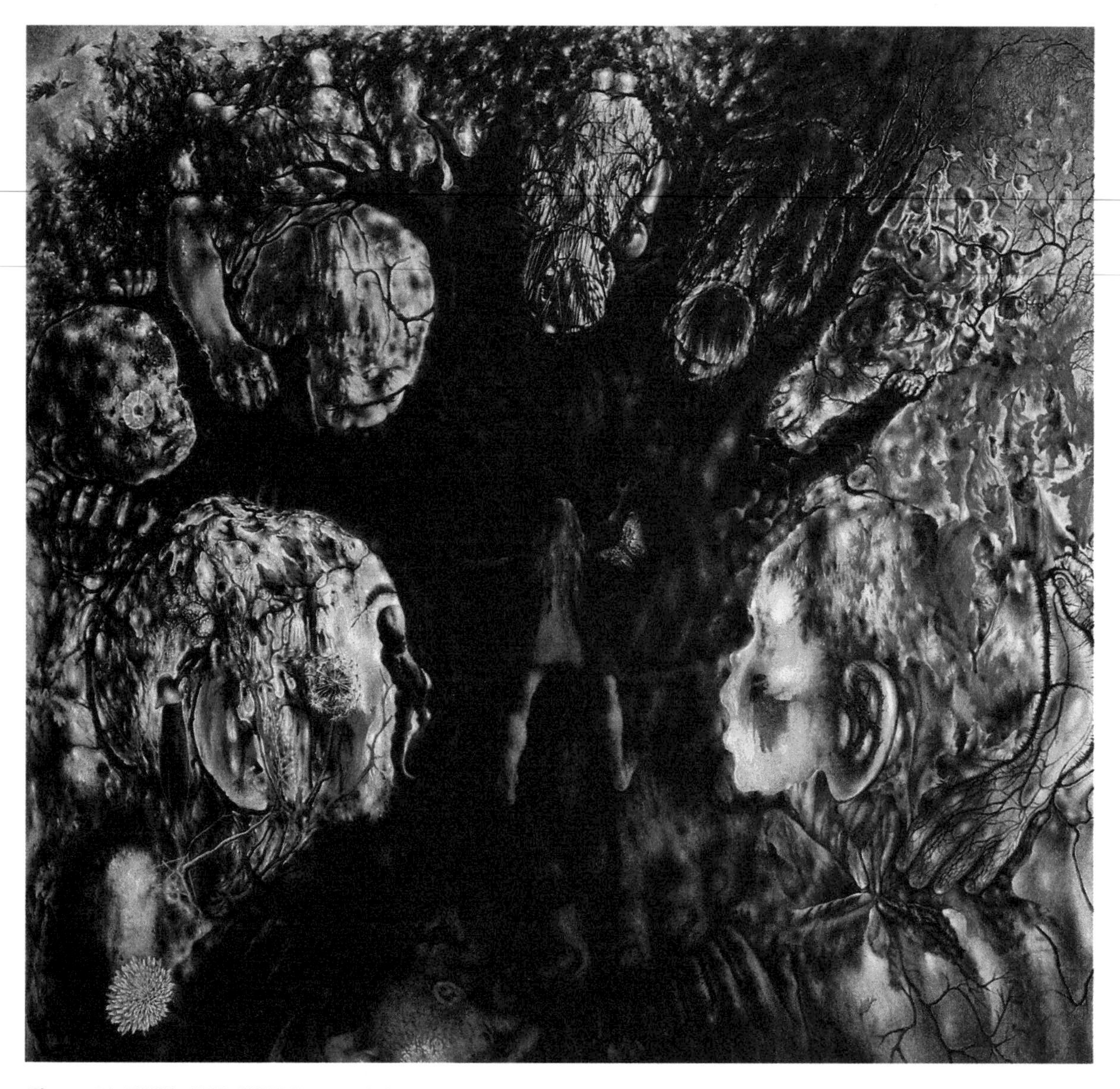

Figure 1.1 ***HIDE-AND-SEEK*** **(1940–42), by Pavel Tchelitchew, captures interplay between the mind and environment that influences the brain's development as well as its architecture. Hidden forms are embedded figures, a delicate test of mental function. Roots, branches and vines suggest neuronal arborization and the ability of such structures to change.**

impulses called action potentials, the beating wings of Cajal's butterflies. Action potentials, which measure about 100 millivolts in amplitude and one millisecond in duration, result from the movement of positively charged sodium ions across the surface membrane from the extracellular fluid into the cell interior, or cytoplasm.

The sodium concentration in the extracellular space is about 10 times the intracellular concentration. The resting membrane maintains a voltage gradient of −70 millivolts; the cytoplasm is negatively charged with respect to the outside. But sodium does not enter rapidly because the resting membrane does not allow these ions easy access.

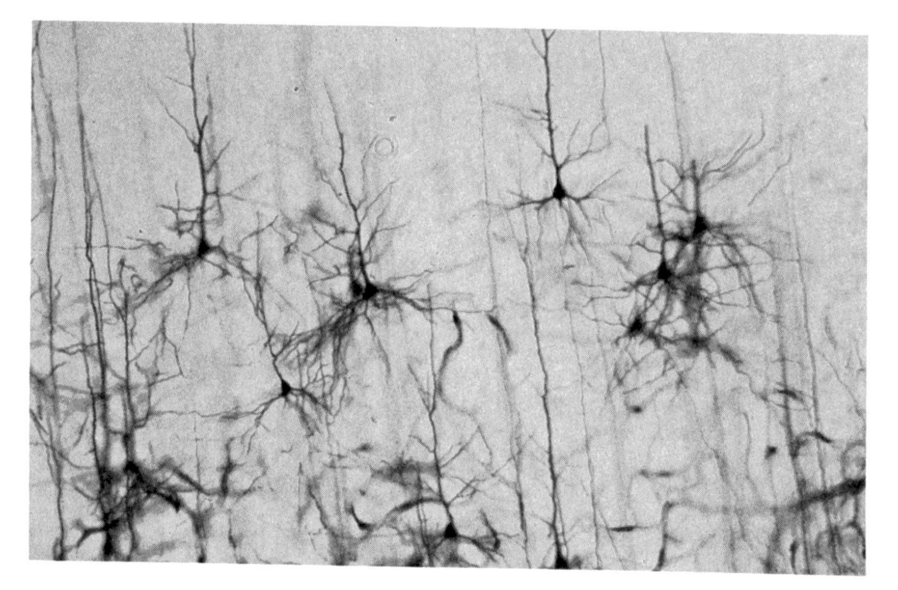

Figure 1.2 NEURONS, revealed by Golgi staining, carry nerve impulses. The cellular architecture of the brain was discovered by Santiago Ramón y Cajal. Janet Robbins in David H. Hubel's laboratory at Harvard Medical School made this preparation.

Physical or chemical stimuli that decrease the voltage gradient, or depolarize the membrane, increase sodium permeability. Sodium influx further depolarizes the membrane, thus increasing sodium permeability even more.

At a critical potential called the threshold, the positive feedback produces a regenerative event that forces the membrane potential to reverse in sign. That is, the inside of the cell becomes positive with respect to the outside. After about one millisecond the sodium permeability declines, and the membrane potential returns to −70 millivolts, its resting value. The sodium permeability mechanism remains refractory for a few milliseconds after each explosion. This limits to 200 per second or less the rate at which action potentials can be generated.

Although axons look like insulated wires, they do not conduct impulses in the same way. They are not good cables: the resistance along the axis is too high and the membrane resistance too low. The positive charge that enters the axon during the action potential is dissipated in one or two millimeters. To travel distances that may reach many centimeters, the action potential must be frequently regenerated along the way. The need to boost repeatedly the current limits the maximum speed at which an impulse travels to about 100 meters per second. That is less than one millionth of the speed at which an electrical signal moves in a copper wire. Thus, action potentials are relatively low frequency, stereotypical signals that are conducted at a snail's pace. Fleeting thoughts must depend on the relative timing of impulses conducted over many axons in parallel and on the thousands of connections made by each one.

The brain is not a syncytium, at least not a simple one. Action potentials cannot jump from one cell to another. Most often, communication between neurons is mediated by chemical transmitters that are released at specialized contacts called synapses. When an action potential arrives at the axon terminal, transmitters are released from small vesicles in which they are packaged into a cleft 20 nanometers in width that separates presynaptic and postsynaptic membranes. Calcium ions enter the nerve terminal during the peak of the action potential. Their movement provides the cue for synchronized exocytosis, the coordinated release of the neurotransmitter molecules.

Once released, transmitters bind to postsynaptic receptors, triggering a change in membrane permeability. The effect is excitatory when the movement of charge brings the membrane closer to the threshold for action-potential generation. It is inhibitory when the membrane is stabilized near its resting value. Each synapse produces only a small effect. To set the intensity (action-potential frequency) of

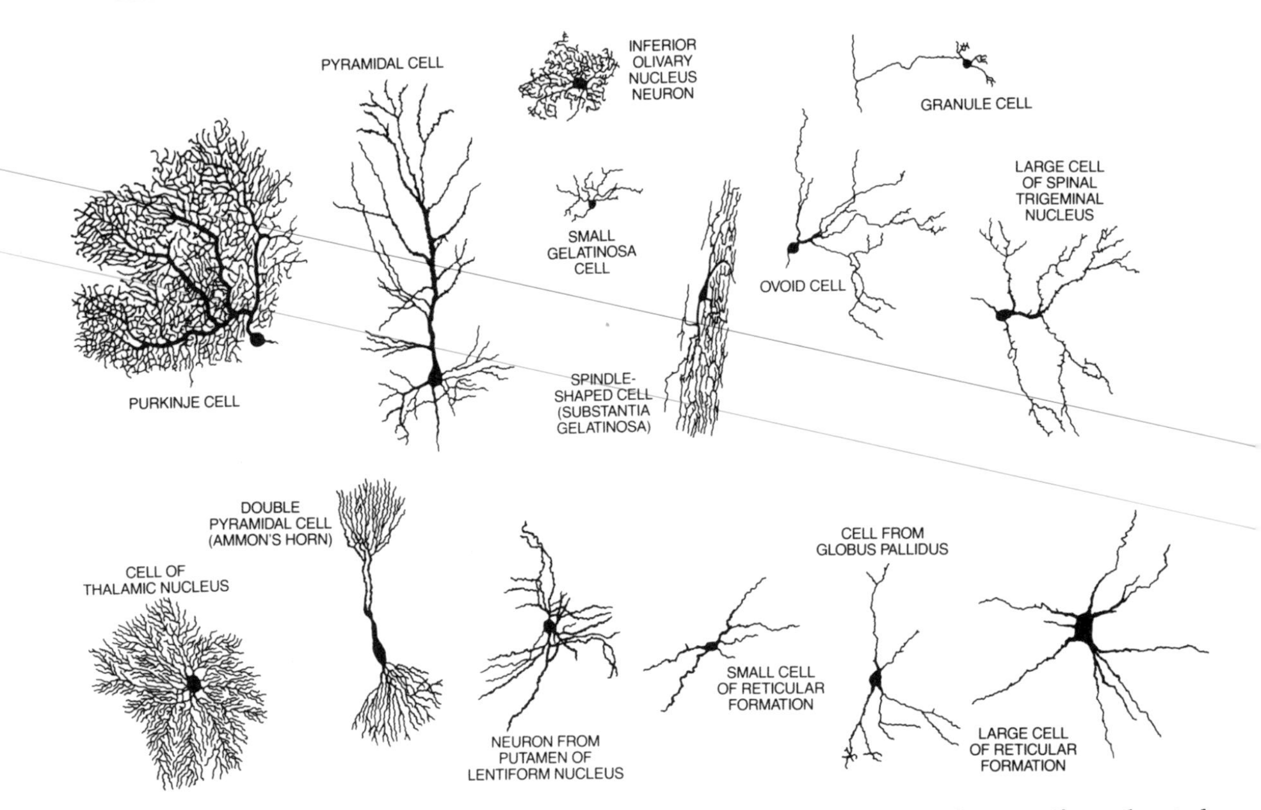

Figure 1.3 STRUCTURAL VARIETY OF NEURONS (shown as tracings from Golgi stains) contributes to the vast capacity of the brain to store, retrieve, use and express information, as well as to experience emotion and control movement.

its output, each neuron must continually integrate up to 1,000 synaptic inputs, which do not add up in a simple linear manner. Each neuron is a sophisticated computer.

Many different kinds of transmitters have been identified in the brain, and this variety has enormous implications for brain function. Since the first neurotransmitter was identified in 1921, the list of candidates has grown at an increasing pace. Fifty is close to the mark. We have learned a great deal about how transmitters are synthesized, how they are released and how they activate receptors in the postsynaptic membrane.

This level of analysis is particularly relevant for psychiatric and neurological disorders that shed light on the workings of the mind (see Chapter 8, "Major Disorders of Mind and Brain," by Elliot S. Gershon and Ronald O. Rieder). For example, drugs that alleviate anxiety, such as Valium, augment the action of gamma-aminobutyric acid (GABA), an important inhibitory transmitter. Antidepressants, such as Prozac, enhance the action of serotonin, an indoleamine with a wide variety of functions. Cocaine facilitates the action of dopamine, whereas certain antipsychotics antagonize this catecholamine. Nicotine activates acetylcholine receptors, which are distributed throughout the cerebral cortex. Further insight into the chemical bases of thinking and behavior depends on obtaining more precise data regarding the sites of action of these potent agents and on the discovery of more selective ligands, molecules that bind to receptors.

The power of the molecule-to-mind approach can be illustrated by recent advances in the pharmacologic treatment of schizophrenia, the most common and the most devastating of all thought disorders.

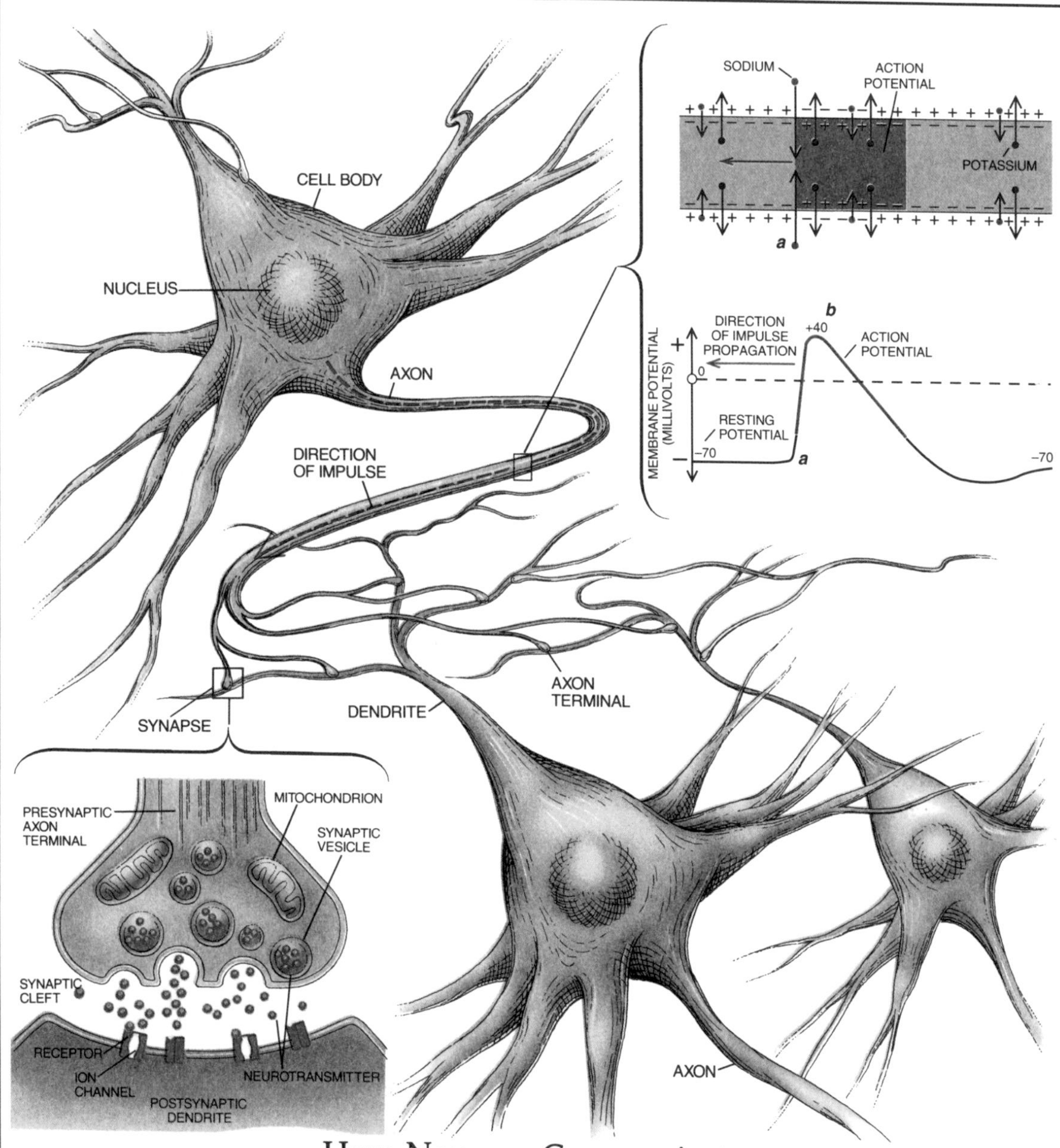

How Neurons Communicate

An excited neuron (*turquoise*) conveys information to other neurons (*purple*) by generating action potentials. These signals propagate like waves down the length of the axon and are converted to chemical signals at synapses. When a neuron is at rest, its external membrane maintains an electrical potential difference of about −70 millivolts and the membrane is more permeable to potassium ions than to sodium ions. When the cell is stimulated, the permeability to sodium increases, leading to an inrush of positive charges (*a*), which triggers an impulse—a momentary reversal (*b*) of the membrane potential. The impulse is conducted away from the cell body (*red arrows*) and when it reaches the axon terminals of the presynaptic neuron, it induces the release of neurotransmitter molecules (*inset at bottom left*).

The classic antipsychotic drugs include the phenothiazines (for example, Thorazine) and the butyrophenones (for example, Haldol). These agents ameliorate hallucinations, delusions, disorganized thinking and inappropriate affect—the "positive" symptoms of schizophrenia that are most evident during acute psychotic episodes. They are not as effective in treating autism and paucity of speech—"negative" symptoms that are prominent during interpsychotic intervals. Moreover, they all produce subtle, abnormal movements when administered to treat acute episodes of illness (hence the name "neuroleptics"). When administered for a long time, they often cause a devastating disorder called tardive dyskinesia. Involuntary and at times incessant writhing movements of the limbs and trunk characterize the disorder, which can persist long after the drug is discontinued.

Why would an agent that affects mental function also produce motor symptoms? The answer lies in the fact that conventional antipsychotics prevent the binding of dopamine to its receptors. To appreciate the importance of this insight, one must know that dopamine-containing nerve cell bodies, gathered deep in the midbrain in a region known as the ventral tegmentum, project their axons widely to the prefrontal cortex as well as to subcortical structures, including the basal ganglia, which are involved in many aspects of motor control. The prefrontal cortex is particularly relevant to schizophrenia because it contains circuits that are active during manipulation of symbolic information and in a type of short-term memory called working memory (see Chapter 6, "Working Memory and the Mind," by Patricia S. Goldman-Rakic). Neurons in this region may form a central processing unit of sorts.

A new drug, clozapine, affects the negative as well as the positive symptoms of schizophrenia. Most important, clozapine does not lead to tardive dyskinesia. The discovery of additional members of the dopamine receptor family may provide the explanation for the unique efficacy and selectivity of this antipsychotic.

Transmitter receptors can be grouped into two large (and growing) superfamilies based on their amino acid sequence and on presumptions about the shape that the molecules assume as part of the cell membrane in which they are embedded. A more detailed receptor classification scheme has emerged. It incorporates molecular architecture as well as the more traditional criteria of ligand binding and function. Based on the added molecular information, one receptor superfamily consists of ion channels, proteins that can form aqueous pores through which ions cross the membrane. They underlie the changes in permeability discussed above. The other superfamily, which includes the dopamine receptors, does not form channels. Instead its members interact with a neighboring membrane protein that cleaves a high-energy phosphate bond from guanosine triphosphate. This process initiates a cascade of biochemical reactions. Such G protein-mediated effects are slow in onset, and they last longer than directly gated receptor responses. It is therefore unlikely that they mediate rapid, point-to-point synaptic transmission in the brain. Rather they modulate the way ion channels respond to stimuli. They set the gain of the system much as the pedals on a piano modulate the action of the keys.

The first dopamine receptor gene was isolated four years ago. The search was based on the presumption that the receptor would resemble other receptors that were known to couple to G proteins. This powerful "homology" screening strategy led in short order to the identification of four more dopamine receptors. One of the recent additions, imaginatively named D_4, has attracted considerable attention. The receptor binds dopamine and clozapine with extraordinarily high affinity. Of equal importance, the D_4 gene is apparently not expressed in the basal ganglia, a finding that may explain the absence of tardive dyskinesia. Precise localization of the D_4 receptor within the prefrontal cortex may reveal the origin of hallucinations or at least a component of the neural machinery that has gone awry in schizophrenia.

The slow rate at which psychoactive drugs work presents a puzzle. Drug receptor interactions are immediate, yet symptoms of schizophrenia, depression and other disorders do not resolve for several weeks. The first consequences of drug binding cannot be the sole explanation for their efficacy. This issue leads to a more general consideration of mechanisms by which the environment might change the brain.

Investigation of dopamine synapses has also provided information about the curse of drug addiction. Cocaine, which binds to and inhibits a protein that transports dopamine away from its site of action, is one of the most powerful reinforcing drugs known. Recent studies point to a neural pathway that may be a target of all addictive substances—amphetamines, nicotine, alcohol and opiates. Within this pathway, the nucleus accumbens, a small subdivision of the basal ganglia, appears to be particularly important. Further studies of neurons in this region will certainly sharpen understanding of drug-seeking behavior. They may reveal mechanisms of motivation in general.

The structural, functional and molecular variety that has been described so far would seem to provide a sufficiently complete basis for mental function. Yet another dimension must be considered: plasticity, the tendency of synapses and neuronal circuits to change as a result of activity. Plasticity weaves the tapestry on which the continuity of mental life depends. Action potentials not only encode information, their metabolic aftereffects alter the circuits over which they are transmitted.

Synaptic plasticity is the basis for the informative connectionist neural models that Geoffrey E. Hinton describes (see Chapter 10, "How Neural Networks Learn from Experience"). More generally, plasticity multiplies the complexity provided by any fixed cast of molecular characters or cellular functions. Hence, it provides an even richer substrate for mental phenomena.

From the brief tour of synaptic biology presented above, you can imagine many ways that synaptic efficacy might be altered. For example, transmitter release can be enhanced by a small increase in the amount of calcium that enters a nerve terminal with each action potential. The probability of postsynaptic receptor activation can be changed, and on a longer time scale, variations in activity can alter the number of functional receptors. Increases or decreases in the number of receptors, which take time to occur, may account for the delayed effect of psychotherapeutic agents. Beyond changes in the function of synapses, activity may alter the number or location of synapses themselves. Axons sprout new endings when their neighbors become silent, and the terminal branches of dendritic arbors are constantly remodeled (see Chapter 9, "Aging Brain, Aging Mind," by Dennis J. Selkoe).

In their discussion of plasticity and learning (see Chapter 4, "The Biological Basis of Learning and Individuality"), Eric R. Kandel and Robert D. Hawkins review evidence that short-term synaptic changes associated with simple forms of learning are accompanied by molecular modification of proteins. One such modification is phosphorylation, the addition or attachment of a phosphate group. Phosphorylation has a profound effect on the function of proteins. It is commonly stimulated by transmitters and drugs that act via G protein-coupled receptors. But proteins are degraded on a time scale that ranges from minutes to days. Maintenance of memories that may last a lifetime requires more stable alterations, such as those associated with persistent changes in gene expression. A recently discovered family of genes called immediate early genes (IEGs), which are activated rapidly by brief bursts of action potentials, may provide a crucial link. As expected of master switches that initiate long-term changes in the brain, IEGs encode transcription factors, proteins that regulate the expression of other genes.

Some evidence has been obtained that impulse activity increases the expression of genes that encode trophic factors, proteins that promote the survival of neurons. The adage "use it or lose it" may soon have a specific biochemical correlate. The actions of each transcription factor and their relevance remain to be determined, however.

Another focus of inquiry into the basis of memory is the phenomenon of long-term potentiation (LTP), a persistent increase in synaptic efficacy that follows brief periods of stimulation. Attention has focused on synapses in the hippocampus because clinical and experimental data have implicated this region of the cortex in forms of memory that require conscious deliberation. At certain synapses in the hippocampus, LTP may last for weeks. At the same junctions, LTP meets the "Hebbian" criterion for learning in that it requires coincident presynaptic and postsynaptic activity. LTP does not occur if the postsynaptic neuron is rendered inactive during the priming, presynaptic stimulation. Donald O. Hebb suggested this relation in his 1949 book *The Organization of Behavior* as a basis for the formation of new neural ensembles during learning. It has been repeated often enough to have achieved the force of law.

Synaptic transmission in the hippocampus is mediated by glutamate, the most common excitatory transmitter in the brain, and LTP of the Hebbian type is blocked by aminophosphonovaleric acid (APV), a selective antagonist of one type of glutamate receptor. APV also diminishes the ability of rats to learn tasks that require spatial cues. This is probably not a coincidence, but it remains to be shown that these observations are causally related. The gene that encodes the APV-sensitive glutamate receptor has been cloned in recent months. We can therefore expect tests in transgenic mice bearing mutated receptors to be conducted in the near future. The work will not be straightforward. The plasticity of the brain and the likelihood that natural selection has provided alternative routes to such an important end may complicate matters.

Although the forces leading to plastic changes in the mature brain are ubiquitous and unrelenting, it is important to emphasize the precision and overall stability of the wiring diagram. We could not sense the environment or move in a coordinated manner, let alone think, if it were otherwise. All studies of higher brain function must take into account the precise way in which neurons are connected to one another.

Pathways in the brain have been traced by means of a variety of molecules that are transported along axons. Such reporter molecules can be visualized once the tissue is properly prepared. Connections have also been traced by fine-tipped microelectrodes positioned close enough to a nerve cell body or an axon to detect the small currents generated as an action potential passes by. Each technique has revealed ordered, topographic maps in the cerebral cortex. The body surface is represented in the postcentral gyrus of the cerebral cortex even though the cortical neurons are three synapses away from sensory receptors in the skin. Likewise, a point-to-point map of the visual world is evident in the primary visual cortex at the occipital pole at the back of the brain. Order is evident at each of the early relays on route to the cortex, and topographic order has also been found in projections from the primary cortices to higher centers.

To appreciate just how precise the wiring diagram can be, we need only consider a fundamental discovery made about 30 years ago by Hubel and Wiesel. They determined that neurons in the primary visual cortex (V1) respond to line segments or edges of a particular orientation rather than to the small spots of light that activate the input neurons in the retina and lateral geniculate nucleus of the thalamus. The response implies that neurons in V1 are connected, via the lateral geniculate nucleus, to retinal ganglion cells that lie along a line of the preferred orientation.

We know the anatomy of the major sensory and motor systems in some detail. In contrast, the pattern of connections within the intervening association cortices and the large subcortical nuclei of the cerebral hemispheres is not clearly defined. Goldman-Rakic's experiments are designed to decipher the wiring diagram of the monkey's prefrontal cortex in order to provide a more complete anatomy of memory. Our lack of information about similar connections in the human brain is glaring. Unlike the molecular building blocks and the functions of individual neurons, it cannot be assumed that the intricacies of cortical connectivity will be conserved in different species. The intricacy of this network, after all, is what distinguishes *Homo sapiens* from all other forms of life. An effort akin to the genome project may be called for.

How does the specificity of synaptic connections come about during development? Carla J. Shatz reviews mechanisms by which axons are guided to their appropriate targets in the visual and other systems (see Chapter 2, "The Developing Brain"). The initial stages of axon outgrowth and pathway selection are thought to occur independently of activity. The genetically determined part of the program is evident in the remarkably complete wiring diagram that forms during embryonic life. But once the advancing tips of the axons arrive in the appropriate region, the choices of particular targets are influenced by nerve impulses originating within the brain or stimulated by events in the world itself. Synapse formation during a critical period of development may depend on a type of competition between axons in which those that are activated appropriately are favored.

Steroid hormones also influence the formation of synapses during early development at least in certain regions of the brain (see Chapter 7, "Sex Differences in the Brain," by Doreen Kimura). Anatomic, physiological and behavioral data indicate that the brains of males and females are not identical.

The pattern of information flow in the brain during the performance of mental tasks cannot easily be determined by anatomic studies of the circuit diagram or by studies of plasticity. Neural correlates of higher mental functions are being sought directly in awake primates trained to perform tasks that require judgment, planning or memory, or all three capacities. This demanding approach requires sophisticated instrumentation, sophisticated experimental design and months of training until the monkey thinks the same thoughts as the investigator. All-night sessions spent listening to amplified action potentials generated by one or a few neurons followed by days of data analysis are the rule. Progress is slow, but important generalizations have emerged.

One of the most important principles is that sensory systems are arranged in a hierarchical manner. That is, neurons respond to increasingly abstract aspects of complex stimuli as the distance—measured in numbers of synapses from the source—grows. The fact that neurons in V1 respond to lines rather than spots makes the case. Another important principle, discussed by Semir Zeki, is that information does not travel along a single pathway. Rather, different features of a single percept are processed in parallel pathways (see Chapter 3, "The Visual Image in Mind and Brain"). A tennis player who wanders to the net from time to time will be alarmed to learn that the movement, color and shape of a tennis ball are processed in different cortical visual centers. Separation of these information streams begins in the retina; they remain segregated in the lateral geniculate nucleus and the primary visual cortex en route to the higher visual centers.

An analogous situation has been found in the auditory system. Mark Konishi and his colleagues at the California Institute of Technology have shown that the localization of sound sources by the barn owl depends on interaural phase and amplitude differences. Phase differences indicate location along the azimuth, whereas amplitude differences signal elevation. Phase and amplitude signals are processed in different pathways through three synaptic relays in the brain. It seems likely that this type of parallel processing characterizes other sensory systems, association cortices and motor pathways as well.

Where is the information reassembled? When does the subject become aware of the approaching ball? The receptive fields of neurons in higher centers are larger than those found in earlier relay stations, so they monitor a larger fraction of the external world. Zeki describes a model that depends on feedback connections from cells with large receptive fields to the cells in the primary visual cortex that have high spatial resolution. Such feedback circuits might coordinate the activity of cells in the primary cortex that have high spatial resolution and cells that respond to more abstract features of the stimulus no matter where it is located. Francis Crick and Christof Koch address the role in visual awareness of a 40-cycle-per-second oscillation in firing rate that is observed throughout the cortex (see Chapter 11, "The Problem of Consciousness"). The oscillations, discovered by Wolf J. Singer and his colleagues at the Max Planck Institute for Brain Research in Frankfurt, may synchronize the firing of neurons that respond to different components of a perceptual scene and hence may be a direct neural correlate of awareness.

Konishi has identified the first neurons in the owl's brain that respond to a combination of interaural phase and amplitude differences but not to either parameter presented alone. These neurons, located deep in the animal's brain in a region called the inferior colliculus, activate a motor program that results in the owl's turning toward the sound source.

In the monkey's visual system, "face cells" located in the inferior temporal sulcus represent perhaps the highest level of abstraction yet identified. These neurons respond to faces but not to other visual stimuli. Similar cells may be present in our own brains. Lesions in the corresponding area of the temporal lobe result in prosopagnosia, a remarkably selective deficit in which the ability to recognize faces is lost. In the zebra finch's auditory system (birds again), a high level of abstraction is evident in neurons found in each male's brain that respond to the complex song of his father but not to pure tones or to the songs of other males of the same species.

How many neurons must change their firing rate to signal a coherent percept or gestalt? The most extreme view holds that one cell may do the job. Is there one face cell per face? Such a supposition seems unlikely on first principles: we lose thousands of neurons every day, so overcommitment to one would be unwise. A more compelling argument comes from recent experiments that have shown face cells to be broadly tuned, responding to faces with similar features rather than to one face alone. The number of neurons that must be activated before recognition emerges is not known, but the data are consistent with a sparse coding rather than global or diffuse activation.

Face cells have their counterparts on the motor side. "Command" neurons have been identified in certain invertebrates that trigger all-or-none, fixed-action patterns, such as stereotypical escape behaviors. Apostolos P. Georgopoulos of Johns Hopkins University has found command neurons of a kind in the monkey's motor cortex (precentral gyrus) that encode the direction of forelimb movement. The firing of these neurons is not associated with the contraction of a particular muscle or with the force of the coordinated movement. Like face cells in the temporal lobe, individual motor cortex neurons are broadly tuned.

The vector obtained by summing the firing frequencies of many neurons is better correlated with the direction of movement than is the activity of any individual cell (see Figure 1.4). The vector becomes evident several milliseconds *before* the appropriate muscles contract and the arm actually moves. It must be a sign of motor planning. The vector is usually derived from less than 100 neurons, so sparse coding may be the rule in the motor cortex as it is in the temporal sulcus.

An important next step at this level of analysis is to produce mental phenomena by focal electrical stimulation. A beginning has been made by William Newsome and his colleagues at Stanford University. They trained monkeys to decide on the direction of movement of dots displayed in random positions on a screen. When the number of dots that showed net movement was set near the threshold for a consistent judgment about the population as a whole, focal stimulation of the V5 region in the cortex influenced the monkey's perceptual judgments.

Strokes and other unfortunate "experiments of nature" have also provided important insights regarding neural correlates of mental phenomena. Antonio R. and Hanna Damasio continue a long tradition of research in their studies of language disorders among neurological patients (see Chapter 5, "Brain and Language"). This work requires careful examination with a battery of tests designed to elicit the most subtle deficits. Here is an example of the pressing need to define the mental phenomena that need to be explained. The Damasios propose the view that language can be considered a three-part system: word formation, concept representation and mediation between the two. If, as they suggest, language has evolved as a tool to compress concepts and communicate them in an efficient manner, a clear view of its functional anatomy brings us to the crux of the mind matter.

The very real experience of phantom limbs cautions against quick acceptance of sparse coding or even of localization as a universal mechanism. Amputated limbs, experiments of nature of another sort, may be experienced as an integral part of the body or "self" [see "Phantom Limbs," by Ronald

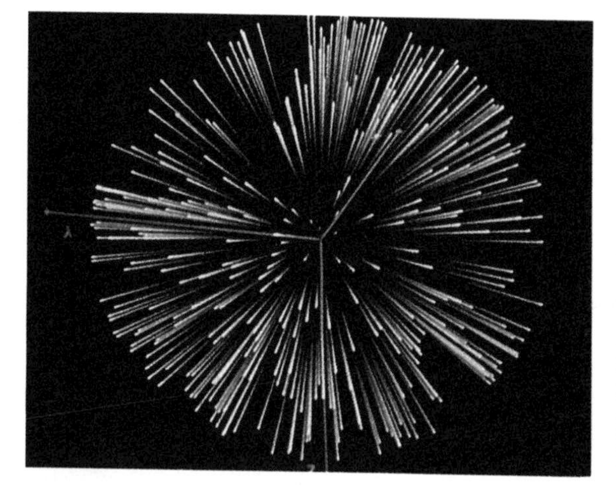

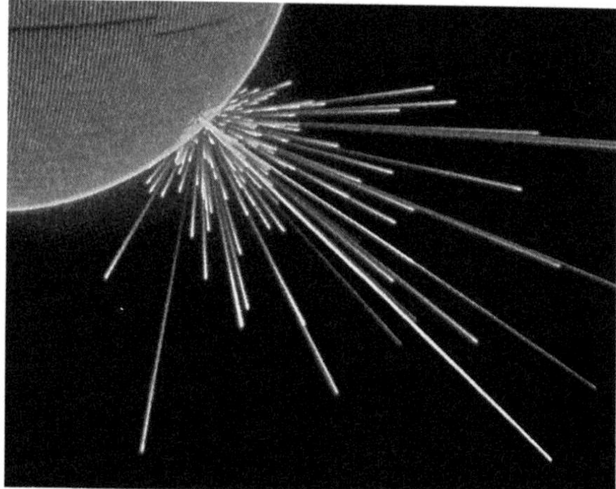

Figure 1.4 INTENT of a monkey to move its arm is revealed by electrical activity of neurons in the motor cortex. Microelectrodes recorded the impulses. Each line represents the rate of firing of individual neurons. Computer diagram at the left shows firing associated with the full range of arm movement. Diagram at the right shows the firing of neurons controlling movement in one direction only (*long yellow line*). The direction of the population vector (*orange line*) is close to that of the actual movement. Apostolos P. Georgopoulos of Johns Hopkins University and his colleagues made the measurements.

Melzack; SCIENTIFIC AMERICAN, April 1992]. A deep and burning pain is a distressing component of the syndrome. It is impossible to find a local area in which such sensations are experienced. Attempts have been made to abolish phantom pain by cutting peripheral nerves, by destroying ascending pathways and by removing sensory regions of the brain. All attempts have failed to eliminate the perception of pain. It may be that the emotional response we call pain requires activation of neurons in widely dispersed regions of the brain.

The future of cognitive neuroscience depends on our ability to study the living human brain. Positron emission tomography (PET) and functional magnetic resonance imaging (MRI) hold great promise in this regard. These noninvasive imaging techniques depend on tight coupling between neuronal activity, energy consumption and regional blood flow. These relations were pointed out by Sir Charles Scott Sherrington in 1890 and later placed on a quantitative basis by Seymour S. Kety and Louis Sokoloff of the National Institute of Mental Health. The brain is never completely at rest. Furthermore, the increases in regional blood flow that MRI and PET detect are not large (they are on the order of 20 to 50 percent). So PET and MRI measurements depend on sophisticated subtraction algorithms that allow one to distinguish the pattern of blood flow during the mental task from the resting, or control, pattern (see Figure 1.5). Assignment of

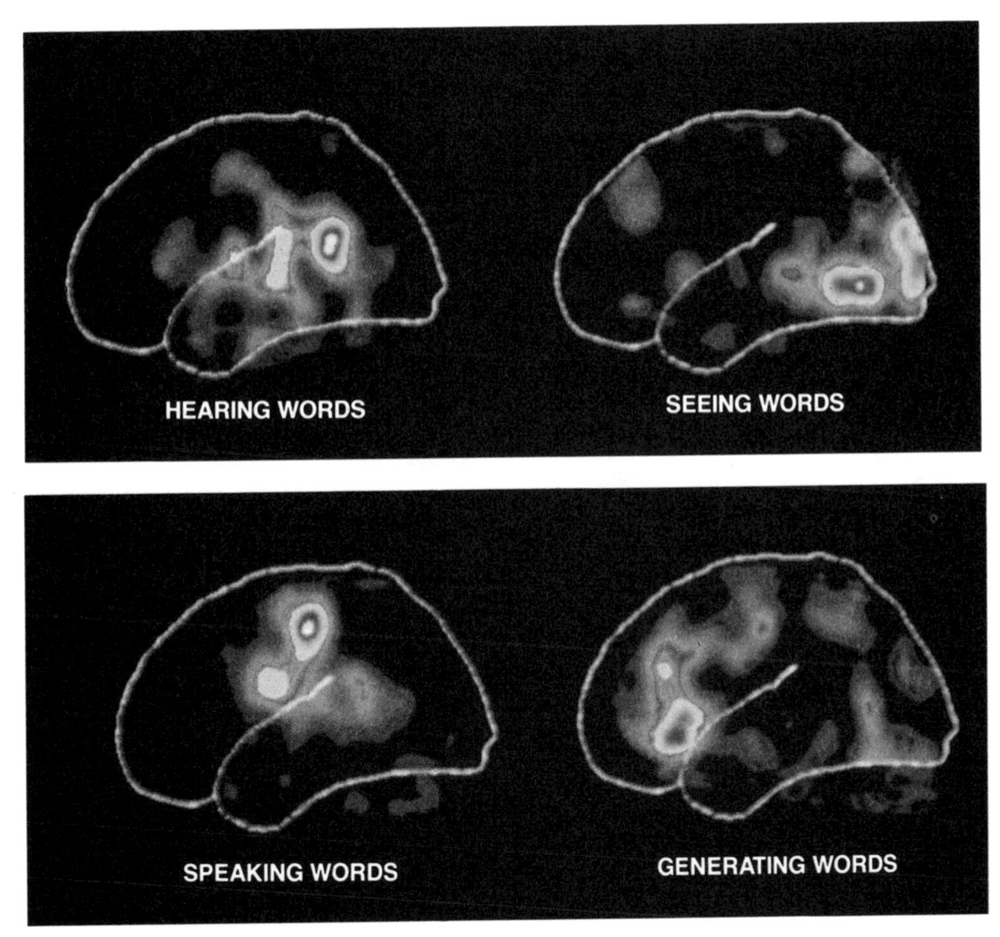

Figure 1.5 PET SCANS show the brain of a human subject performing a series of intellectual tasks related to words. The positron emission tomographic technique reveals that blood flow in the brain shifts to different locations, depending on which task is being performed. Marcus Raichle of the Washington University School of Medicine made the images.

the changes in blood flow to specific structures depends on accurately superimposing the computed images on precise anatomic maps.

At present, neither technique provides the spatial resolution to visualize single cortical columns. Moreover, the slow temporal resolution of both imaging techniques demands that mental tasks be repeated over and over again during the recording session. Technical advances, especially related to rapid MRI scanning, are sure to follow. Even with the current limitations, the advantages of working with humans, who can think on command, are overwhelming.

In sum, we can expect advances at an increasing rate on all levels of investigation relevant to the mind. We will soon know exactly how many transmitters and transmitter receptors there are in the brain and where each one is concentrated. We will also have a more complete picture of neurotransmitter actions, including multiple interactions of jointly released modulators. And we will learn much more about molecules that affect neuronal differentiation and degeneration. Molecules of the mind are not unique. Many of the neurotransmitters are common amino acids found throughout the body. Likewise, no new principles or molecules specific to the brain have emerged in studies of hormone regulation or of trophic factors that influence the survival and differentiation of neurons. The great challenge, then, is to determine how these molecules modulate the functional wiring diagram of the brain and how this functional nerve net gives rise to mental phenomena.

Ultimately, it will be essential to specify what exactly it means to say that mental events are correlated with electrical signals. Certainly, there is a need for theory at this level of analysis, and as emphasized by Crick and Koch, this effort has become one of the most exciting aspects of cognitive neuroscience.

Is the mind an emergent property of the brain's electrical and metabolic activity? An emergent property is one that cannot be accounted for solely by considering the component parts one at a time. For example, the heart beats because its pacemaker depends on the influx and efflux of certain ions. But the automaticity cannot be understood without considering the magnitude and kinetics of all the fluxes together. Once that is accomplished, what is left to explain in physiological terms? In an analogous manner, biological explanations of mental events may become evident once the component neural functions are more clearly defined. We will then have a more appropriate vocabulary for describing the emergent mind.

The Developing Brain

During fetal development, the foundations of the mind are laid as billions of neurons form appropriate connections and patterns. Neural activity and stimulation are crucial in completing this process.

• • •

Carla J. Shatz

An adult human brain has more than 100 billion neurons. They are specifically and intricately connected with one another in ways that make possible memory, vision, learning, thought, consciousness and other properties of the mind. One of the most remarkable features of the adult nervous system is the precision of this wiring. No aspect of the complicated structure, it would appear, has been left to chance. The achievement of such complexity is even more astounding when one considers that during the first few weeks after fertilization many of the sense organs are not even connected to the embryonic processing centers of the brain. During fetal development (see Figure 2.1), neurons must be generated in the right quantity and location. The axons that propagate from them must select the correct pathway to their target and finally make the right connection.

How do such precise neural links form? One idea holds that the brain wires itself as the fetus develops, in a manner analogous to the way a computer is manufactured; that is, the chips and components are assembled and connected according to a preset circuit diagram. According to this analogy, a flip of a biological switch at some point in prenatal life turns on the computer. This notion would imply that the brain's entire structure is recorded in a set of biological blueprints—presumably DNA—and that the organ begins to work only after the wiring is essentially complete.

Research during the past decade shows that the biology of brain development follows very different rules. The neural connections elaborate themselves from an immature pattern of wiring that only grossly approximates the adult pattern. Although humans are born with almost all the neurons they will ever have, the mass of the brain at birth is only about one fourth that of the adult brain. The brain becomes bigger because neurons grow in size, and the number of axons and dendrites as well as the extent of their connections increases.

Workers who have studied the development of the brain have found that to achieve the precision of the adult pattern, neural function is necessary: the brain must be stimulated in some fashion. Indeed, several observations during the past few decades have shown that babies who spent most of their first year of life lying in their cribs developed abnormally slowly. Some of these infants could not sit up at 21 months of age, and fewer than 15 percent could walk by about the age of three. Children must be stimulated—through touch, speech and images

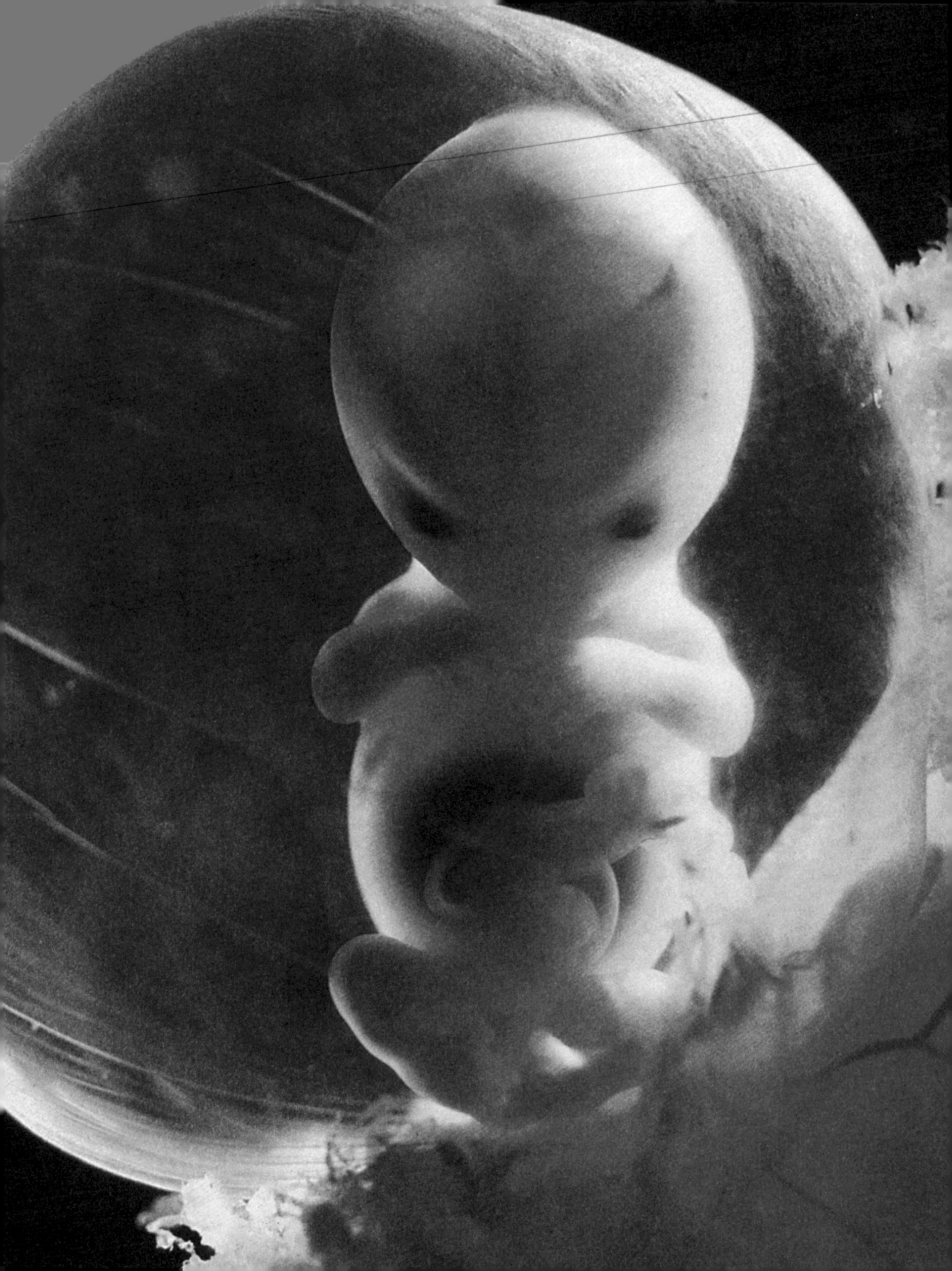

—to develop fully. Based in part on such observations, some people favor enriched environments for young children, in the hopes of enhancing development. Yet current studies provide no clear evidence that such extra stimulation is helpful.

Much research remains to be done before anyone can conclusively determine the types of sensory input that encourage the formation of particular neural connections in newborns. As a first step toward understanding the process, neurobiologists have focused on the development of the visual system in other animals, especially during the neonatal stages. It is easy under the conditions that prevail at that stage to control visual experience and observe behavioral response to small changes. Furthermore, the mammalian eye differs little from species to species. Another physiological fact makes the visual system a productive object of study: its neurons are essentially the same as neurons in other parts of the brain. For these reasons, the results of such studies are very likely to be applicable to the human nervous system as well.

But perhaps the most important advantage is that in the visual system, investigators can accurately correlate function with structure and identify the pathway from external stimulus to physiological response. The response begins when the rods and cones of the retina transform light into neural signals. These cells send the signals to the retinal interneurons, which relay them to the output neurons of the retina, called the retinal ganglion cells. The axons of the retinal ganglion cells (which make up the optic nerve) connect to a relay structure within the brain known as the lateral geniculate nucleus. The cells of the lateral geniculate nucleus then send the visual information to specific neurons located in what is called layer 4 of the (six-layer) primary visual cortex. This cortical region occupies the occipital lobe in each cerebral hemisphere (see Figure 2.2).

Within the lateral geniculate nucleus, retinal ganglion cell axons from each eye are strictly segregated: the axons of one eye alternate with those from the other and thus form a series of eye-specific layers. The axons from the lateral geniculate nucleus in turn terminate in restricted patches within cortical layer 4. The patches corresponding to each eye interdigitate with one another to form structures termed ocular dominance columns.

To establish such a network during development, axons must grow long distances, because the target structures form in different regions. The retinal ganglion cells are generated within the eye. The lateral geniculate neurons take shape in an embryonic structure known as the diencephalon, which will form the thalamus and hypothalamus. The layer 4 cells are created in another protoorgan called the telencephalon, which later develops into the cerebral cortex. From the beginning of fetal development, these three structures are many cell-body diameters distant from one another. Yet after identifying one or the other of these targets, the axons reach it and array themselves in the correct topographic fashion—that is, cells located near one another in one structure map their axons to the correct neighboring cells within the target.

This developmental process can be compared with the problem of stringing telephone lines between particular homes located within specific cities. For instance, to string wires between Boston and New York, one must bypass several cities, including Providence, Hartford, New Haven and Stamford. Once in New York, the lines must be directed to the correct borough (target) and then to the correct street address (topographic location).

Corey Goodman of the University of California at Berkeley and Thomas Jessel of Columbia University have demonstrated that in most instances, axons immediately recognize and grow along the correct pathway and select the correct target in a highly precise manner. A kind of "molecular sensing" is thought to guide growing axons. The axons have specialized tips, called growth cones, that can recognize the proper pathways. They do so by sensing a variety of specific molecules laid out on the surface of, or even released from, cells located along the pathway. The target itself may also release the necessary molecular cues. Removing these cues (by genetic or surgical manipulation) can cause the axons to grow aimlessly. But once axons have arrived at their targets, they still need to select the correct address. Unlike pathway and target selection, address selection is not direct. In fact, it involves the correction of many initial errors.

Figure 2.1 SEVEN-WEEK-OLD HUMAN FETUS is about an inch long. Eyes and limbs are visible, and the emerging brain is apparent. Stimulation is needed to complete development, a process that for many neural systems continues into neonatal life.

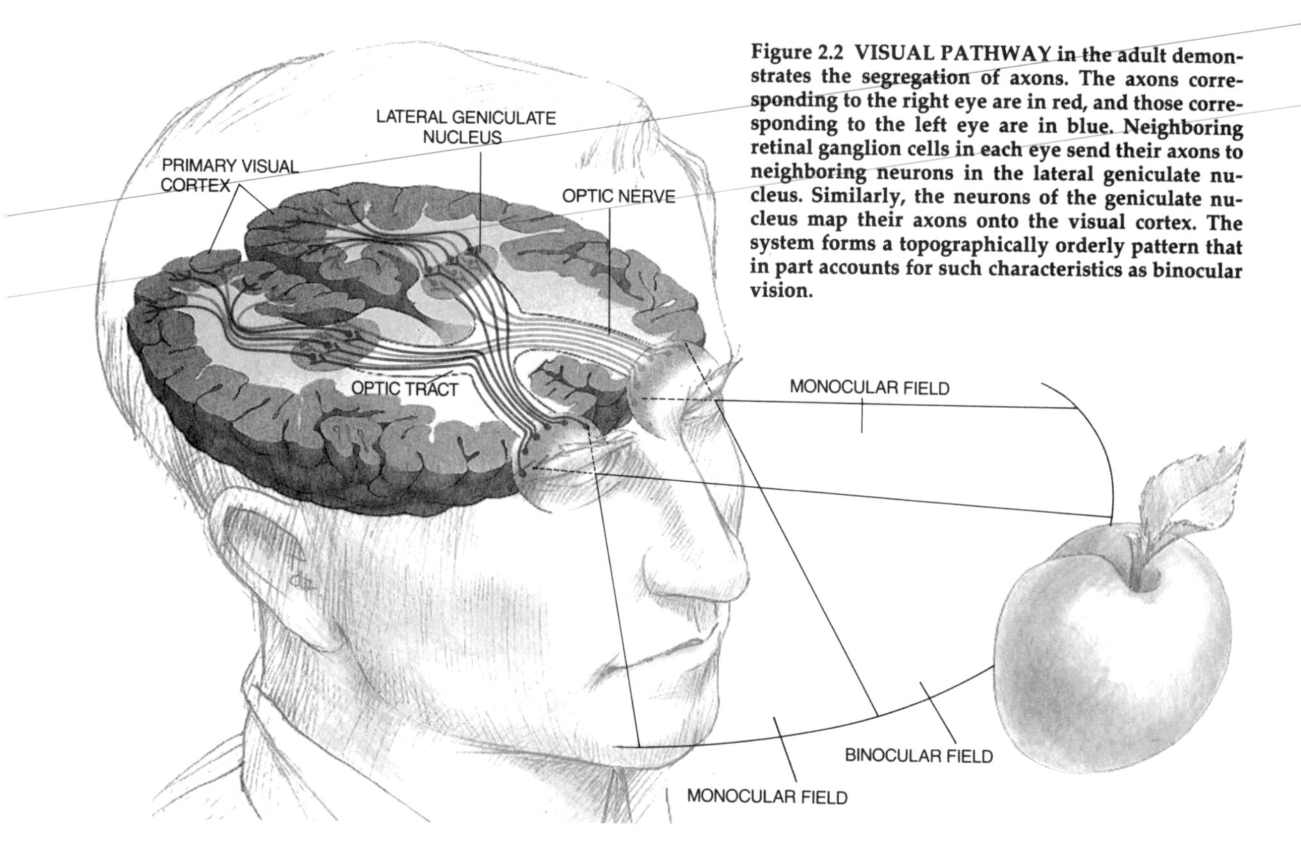

Figure 2.2 VISUAL PATHWAY in the adult demonstrates the segregation of axons. The axons corresponding to the right eye are in red, and those corresponding to the left eye are in blue. Neighboring retinal ganglion cells in each eye send their axons to neighboring neurons in the lateral geniculate nucleus. Similarly, the neurons of the geniculate nucleus map their axons onto the visual cortex. The system forms a topographically orderly pattern that in part accounts for such characteristics as binocular vision.

The first hint that address selection is not precise came from experiments using radioactive tracers. Injections of these tracers at successively later times in fetal development outline the course and pattern of axonal projections. Such studies have also shown that structures emerge at different times in development, which can further complicate address selection.

For instance, Pasko Rakic of Yale University has shown that in the visual pathway in monkeys, the connections between the retina and the lateral geniculate nucleus appear first, followed by those between the lateral geniculate nucleus and layer 4 of the visual cortex. Other studies found that in cats and primates (including humans), the lateral geniculate nucleus layers develop during the prenatal period, before the rods and cones of the retina have formed (and thus before vision is even possible). When Simon LeVay, Michael P. Stryker and I were postdoctoral fellows at Harvard Medical School, we found that at birth, layer 4 columns in cats do not even exist in the visual cortex (see Figure 2.3). I subsequently determined that even earlier, in fetal life, the cat has no layers in the lateral geniculate nucleus. These important visual structures emerge only gradually and at separate stages.

The functional properties of neurons, like their structural architecture, do not attain their specificity until later in life. Microelectrode recordings from the visual cortex of newborn cats and monkeys reveal that the majority of layer 4 neurons respond equally well to visual stimulation of either eye. In the adult, each neuron in layer 4 responds primarily if not exclusively to stimulation of one eye only. This finding implies that during the process of address selection, the axons must correct their early "mistakes" by removing the inputs from the "inappropriate" eye.

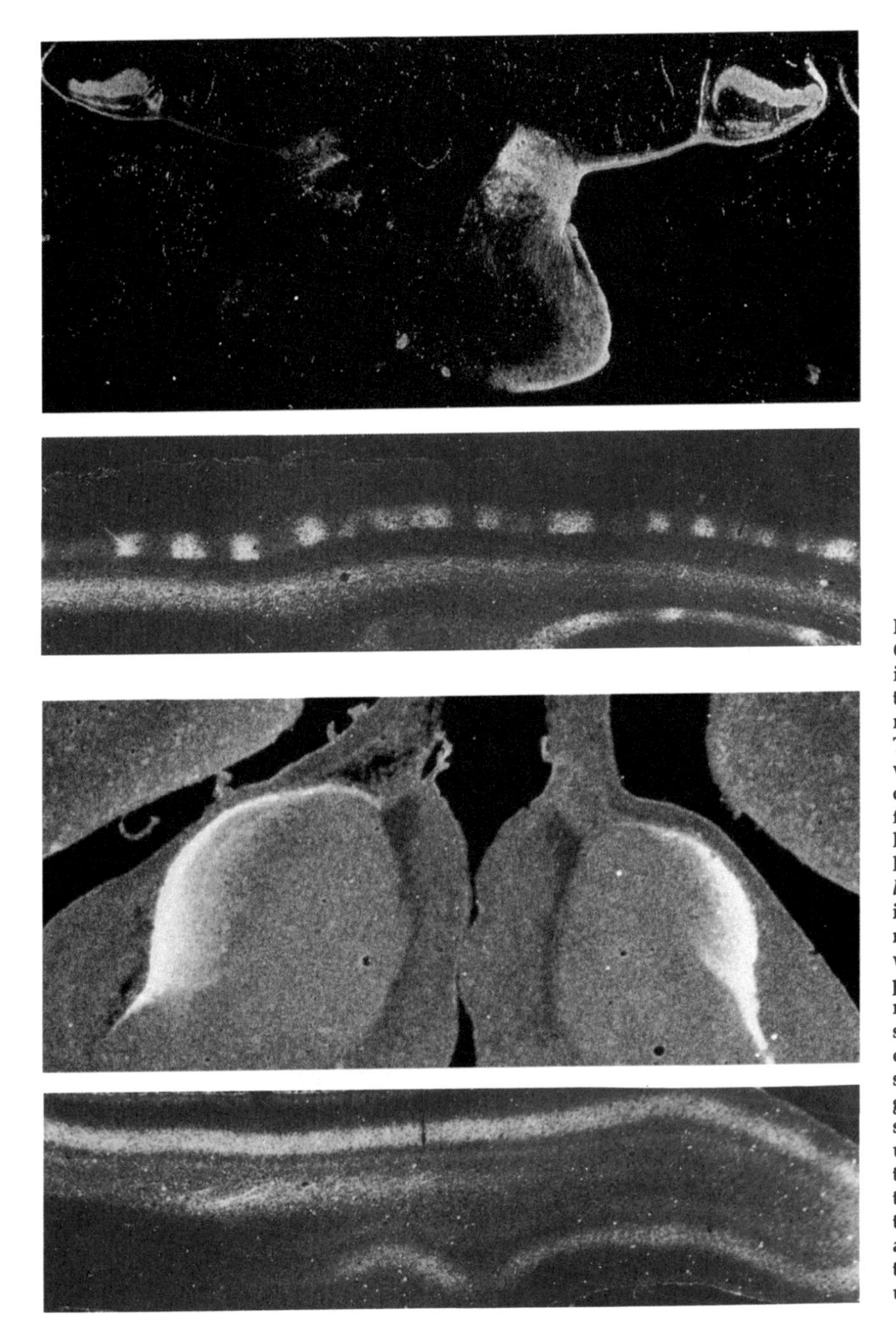

Figure 2.3 NEURAL DEVELOPMENT of the visual system is revealed here by a radioactive tracer injected into the vitreous humor of the left eye. The images correspond to a top view of the visual system of a cat. Only areas receiving input from the injected eye become labeled (*white areas*). In the lateral geniculate nucleus (*top left*), most of the left eye's input ends in layers in the right nucleus, although the wiring of the visual pathway places some label in the left nucleus. Similar segregation is seen in the ocular dominance columns in layer 4 of the visual cortex (*bottom left*); the gaps represent regions corresponding to axons from the uninjected eye. The adult patterns are in distinct contrast to their immature forms, shown to the right. The immature axons have yet to segregate: the label is uniformly distributed.

In 1983 my colleague Peter A. Kirkwood and I found further evidence that axons must fine-tune their connections. It came from our work on the brains of six-week-old cat fetuses (the gestation period of the cat is about nine weeks). We removed a significant portion of the visual pathway—from the ganglion cells in both eyes to the lateral geniculate nucleus—and placed it in vitro in a special life-support chamber. (Inserting microelectrodes in a fetus is extremely difficult.) The device kept the cells alive for about 24 hours. Next we applied electrical pulses to the two optic nerves to stimulate the ganglion cell axons and make them fire action potentials, or nerve signals. We found that neurons in the lateral geniculate nucleus responded to the ganglion cells and, indeed, received inputs from both eyes. In the adult the layers respond only to stimulation of the appropriate eye.

The eventual emergence of discretely functioning neural domains (such as the layers and ocular dominance columns) indicates that axons do manage to correct their mistakes during address selection. The selection process itself depends on the branching pattern of individual axons. In 1986 David W. Sretavan, then a doctoral student in my laboratory, was able to examine the process in some detail. Experimenting with fetal cats, he selectively labeled single retinal ganglion cell axons in their entirety—from the cell body in the retina to their tips within the lateral geniculate nucleus—at successively later stages.

He found that at the earliest times in development, when ganglion cell axons have just arrived within the lateral geniculate nucleus (after about five weeks of gestation), the axons assume a very simple sticklike shape and are tipped with a growth cone. A few days later the axons arising from both eyes acquire a "hairy" appearance: they have short side branches along their entire length.

The presence of side branches at this age implies that the inputs from both eyes mix with one another. In other words, the neural regions have yet to take on the adult structure, in which each eye has its own specific regions. As development continues, the axons sprout elaborate terminal branches and lose their side branches. Soon individual axons from each eye have highly branched terminals that are restricted to the appropriate layer. Axons from one eye that traverse territory belonging to those from the other eye are smooth and unbranched (see Figure 2.4).

The sequence of developmental changes in the branching patterns shows that the adult pattern of

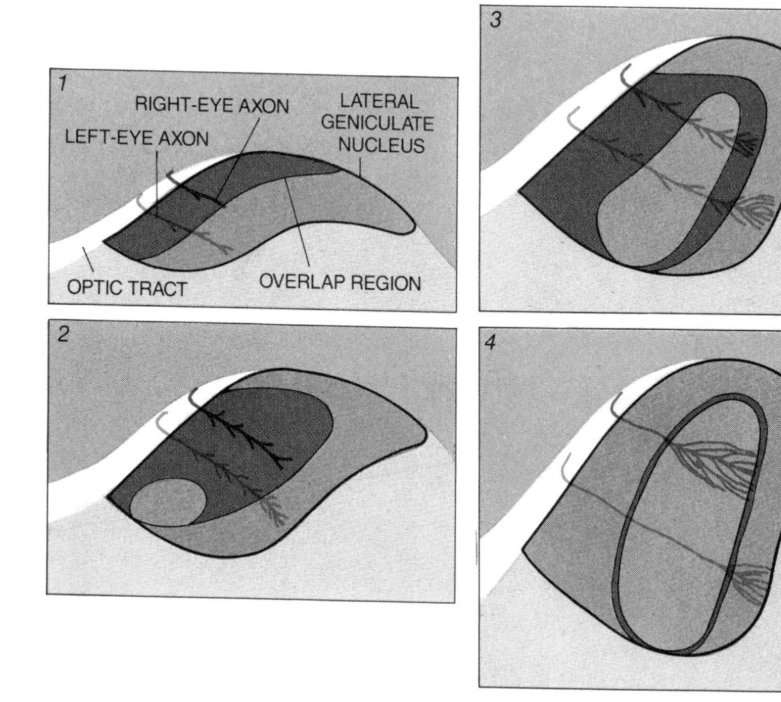

Figure 2.4 AXONAL REMODELING in the lateral geniculate nucleus occurs largely before birth. At the earliest times in development (*1*), the axons from the left eye and right eye are simple and tipped with growth cones. The shaded region represents the intermixing of inputs from both eyes. After further development (*2*), the axons grow many side branches. The axons soon begin to lose some side branches and start to extend elaborate terminal branches (*3*). Eventually these branches occupy the appropriate territory to form eye-specific layers (*4*).

connections emerges as axons remodel by the selective withdrawal and growth of different branches. Axons apparently grow to many different addresses within their target structures and then somehow eliminate addressing errors.

One possible explanation for axonal remodeling is that specific molecular cues are arrayed on the surface of the target cells. Although this idea might seem conceptually attractive, it was very little experimental support. An alternative explanation appears to be stronger. It holds that all target neurons are fair game. Then, some kind of competition between inputs would lead to formation of specific functional areas.

An important clue concerning the nature of the competitive interactions between axons for target neurons has come from the experiments of David H. Hubel of Harvard Medical School and Torsten N. Wiesel of the Rockefeller University. In the 1970s, when both workers were at Harvard, they studied the formation of childhood cataracts. Clinical observations indicated that if the condition is not treated promptly, it can lead to permanent blindness in the obstructed eye. To emulate the effect, Hubel and Wiesel closed the eyelids of newborn cats. They discovered that even a week of sightlessness can alter the formation of ocular dominance columns. The axons from the lateral geniculate nucleus representing the closed eye occupy smaller than normal patches within layer 4 of the cortex. The axons of the open eye occupy larger than normal patches.

The workers also showed that the effects are restricted to a critical period. Cataracts, when they occur in adulthood and are subsequently corrected by surgery, do not cause lasting blindness. Apparently the critical period has ended long ago, and so the brain's wiring cannot be affected.

These observations suggest that the ocular dominance columns form as a consequence of use. The axons of the lateral geniculate nucleus from each eye somehow compete for common territory in layer 4. When use is equal, the columns in the two eyes are identical; unequal use leads to unequal allotment of territory claimed in layer 4.

How is use translated into these lasting anatomic consequences? In the visual system, use consists of the action potentials generated each time a visual stimulus is converted into a neural signal and is carried by the ganglion cell axons into the brain. Perhaps the effects of eye closure on the development of ocular dominance columns occur because there are fewer action potentials coming from the closed eye. If that is the case, blockage of all action potentials during the critical period of postnatal life should prevent axons from both eyes from fashioning the correct patterns and lead to abnormal development in the visual cortex. Stryker and William Harris, then a postdoctoral fellow at Harvard, obtained this result when they used the drug tetrodotoxin to block retinal ganglion cell action potentials (see boxed figure "Development and Neural Function"). They found that the ocular dominance columns in layer 4 failed to appear (the layers in the lateral geniculate nucleus were unaffected because they had already formed in utero).

Nevertheless, action potentials by themselves are not sufficient to create the segregated patterns in the cortex. Neural activity cannot be random. Instead it must be defined, both temporally and spatially, and must occur in the presence of special kinds of synapses. Stryker and his associate Sheri Strickland, who are both at the University of California at San Francisco, have shown that simultaneous, artificial stimulation of all the axons in the optic nerves can prevent the segregation of axons from the lateral geniculate nucleus into ocular dominance columns within layer 4. Although this result resembles that achieved with tetrodotoxin, an important difference exists. Here ganglion cell action potentials are present—but all at the same time. Segregation to form the columns in the visual cortex, on the other hand, proceeds when the two nerves are stimulated asynchronously.

In a sense, then, cells that fire together wire together. The timing of action-potential activity is critical in determining which synaptic connections are strengthened and retained and which are weakened and eliminated. Under normal circumstances, vision itself acts to correlate the activity of neighboring retinal ganglion cells, because the cells receive inputs from the same parts of the visual world.

What is the synaptic mechanism that strengthens or weakens the connections? As long ago as 1949, Donald O. Hebb of McGill University proposed the existence of special synapses that could execute the task. The signal strength in each synapses would increase whenever activities in a presynaptic cell (the cell supplying the synaptic input) and in a postsynaptic cell (the cell receiving the input) coincide. Clear evidence showing that such "Hebb synapses" exist comes from studies of the phenomenon of long-term potentiation in the hippocampus. Re-

Development and Neural Function

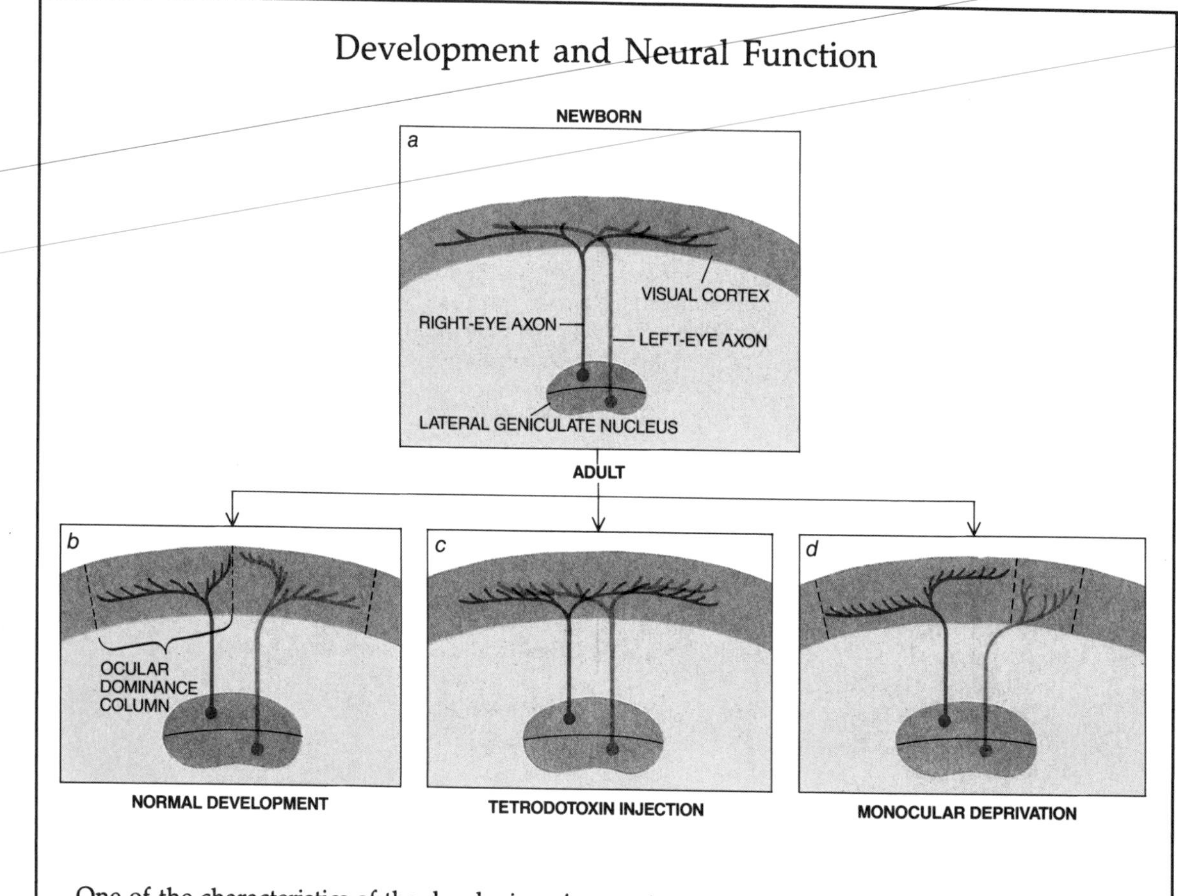

One of the characteristics of the developing visual system is segregation of inputs: each eye adopts its own territory in the visual cortex. The process, however, can be completed only if the neurons are stimulated. In experiments with cat eyes, for example, the axons of the left eye and of the right eye overlap in layer 4 of the visual cortex at birth (*a*). Visual stimuli will cause the axons to separate and form ocular dominance columns in the cortex (*b*). Such normal development can be blocked with injections of tetrodotoxin; as a result, the axons never segregate, and the ocular dominance columns fail to emerge (*c*). Another way to perturb development is to keep one eye closed, depriving it of stimulation. The axons of the open eye then take over more than their fair share of territory in the cortex (*d*).

searchers found that the pairing of presynaptic and postsynaptic activity in the hippocampus can cause incremental increases in the strength of synaptic transmission between the paired cells. The strengthened state can last from hours to days.

Such synapses are now thought to be essential in memory and learning (see Chapter 4, "The Biological Basis of Learning and Individuality," by Eric R. Kandel and Robert D. Hawkins). Studies by Wolf Singer and his colleagues at the Max Planck Institute for Brain Research in Frankfurt and by Yves Fregnac and his colleagues at the University of Paris also suggest that Hebb synapses are present in the visual cortex during the critical period, although their properties are not well understood.

Just how coincident activity causes long-lasting

changes in transmission is not known. There is general agreement among researchers that the postsynaptic cell must somehow detect the coincidence in the incoming presynaptic activity and in turn send a signal back to all concurrently active presynaptic inputs. But this cannot be the whole story. During the formation of the ocular dominance columns, inputs that are not active at the same time are weakened and eliminated.

Consequently, one must also propose the existence of a mechanism for activity-dependent synaptic weakening. This weakening—a kind of long-term depression—would occur when presynaptic action potentials do not accompany postsynaptic activity. Synapses that have this special property (opposite to that of Hebb synapses) have been found in the hippocampus and cerebellum. The results of the Stryker and Strickland experiments suggest that such synapses are very likely to exist in the visual cortex as well.

A strongly similar process of axonal remodeling operates as motor neurons in the spinal cord connect with their target muscles. In the adult, each muscle fiber receives input from only one motor neuron. But after motor neurons make the first contacts with the muscle fibers, each muscle fiber receives inputs from many motor neurons. Then, just as in the visual system, some inputs are eliminated, giving rise to the adult pattern of connectivity. Studies have shown that the process of elimination requires specific temporal patterns of action-potential activity generated by the motor neurons.

The requirement for specific spatial and temporal patterns of neuronal activity might be likened to a process whereby telephone calls are placed from addresses in one city (the lateral geniculate nucleus in the visual system) to those in the next city (the visual cortex) to verify that connections have been made at the correct locations. When two near neighbors in the lateral geniculate nucleus simultaneously call neighboring addresses in the cortex, the telephones in both those homes will ring. The concurrent ringing verifies that relations between neighbors have been preserved during the wiring process.

If, however, one of the neighbors in the lateral geniculate nucleus mistakenly makes connections with very distant parts of layer 4 or with parts that receive input from the other eye, the called telephone will rarely if ever ring simultaneously with those of its neighbors. This dissonance would lead to the weakening and ultimate removal of that connection.

The research cited thus far has explored the remodeling of connections after the animal can move or see. But what about earlier in development? Can mechanisms of axonal remodeling operate even before the brain can respond to stimulation from the external world? My colleagues and I thought the formation of layers in the lateral geniculate nucleus in the cat might be a good place to address this question. After all, during the relevant developmental period, rods and cones have not yet emerged. Can the layers develop their specific territories for each eye even though vision cannot yet generate action-potential activity?

We reasoned that if activity is necessary at these early times, it must somehow be generated spontaneously within the retina, perhaps by the ganglion cells themselves. If so, the firing of retinal ganglion cells might be contributing to layer construction, because all the synaptic machinery necessary for competition is present. It should be possible to prevent the formation of the eye-specific layers by blocking action-potential activity from the eyes to the lateral geniculate nucleus.

Waynesburg College Library
Waynesburg, PA 15370

To hinder activity during fetal development, Sretavan and I, in collaboration with Stryker, implanted special minipumps containing tetrodotoxin in utero just before the lateral geniculate nucleus layers normally begin to form in the cat (at about six weeks of fetal development). After two weeks of infusion, we assessed the effects on the formation of layers. Much to our satisfaction, the results of these in utero infusion experiments showed clearly that the eye-specific layers do not appear in the presence of tetrodotoxin. Moreover, by examining the branching patterns of individual ganglion cell axons after the treatment, we reassured ourselves that tetrodotoxin did not simply stunt normal growth.

In fact, the branching patterns of these axons were very striking. Unlike normal axons at the comparable age, the tetrodotoxin-treated axons did not have highly restricted terminal branches. Rather they had many branches along the entire length of the axon. It was as if, without action-potential activity, the information necessary to withdraw side branches and elaborate the terminal branches was missing.

In 1988, at about the same time these experiments were completed, Lucia Galli-Resta and Lamberto

Waynesburg College Library
Waynesburg, PA 15370

Maffei of the University of Pisa achieved the extraordinary technical feat of actually recording signals from fetal ganglion cells in utero. They demonstrated directly that retinal ganglion cells can indeed spontaneously generate bursts of action potentials in the darkness of the developing eye. This observation, taken together with our experiment, strongly suggests that action-potential activity is not only present but also necessary for the ganglion cell axons from the two eyes to segregate and form the eye-specific layers.

Still, there must be constraints on the spatial and temporal patterning of ganglion cell activity. If the cells fired randomly, the mechanism of correlation-based, activity-dependent sorting could not operate. Furthermore, neighboring ganglion cells in each eye somehow ought to fire in near synchrony with one another, and the firing of cells in the two eyes, taken together, should be asynchronous. In addition, the synapses between retinal ganglion cell axons and neurons of the lateral geniculate nucleus should resemble Hebb synapses in their function: they should be able to detect correlations in the firing of axons and strengthen accordingly.

We realized that to search for such patterns of spontaneous firing, it would be necessary to monitor simultaneously the action-potential activity of many ganglion cells in the developing retina. In addition, the observation had to take place as the eye-specific layers were developing. A major technical advance permitted us to achieve this goal. In 1988 Jerome Pine and his colleagues at the California Institute of Technology, among them doctoral student Markus Meister, invented a special multielectrode recording device. It consisted of 61 recording electrodes arranged as a flat, hexagonal array. Each electrode can detect action potentials generated in one to several cells. When Meister arrived at Stanford University to continue postdoctoral work with Denis Baylor, we began a collaboration to see whether the electrode array could be used to detect the spontaneous firing of fetal retinal ganglion cells (see Figure 2.5).

In these experiments, it was necessary to remove the entire retina from the fetal eye and place it, ganglion-cell-side down, on the array. (It is technically impossible to put the electrode array itself into the eye in utero.) Rachel Wong, a postdoctoral fel-

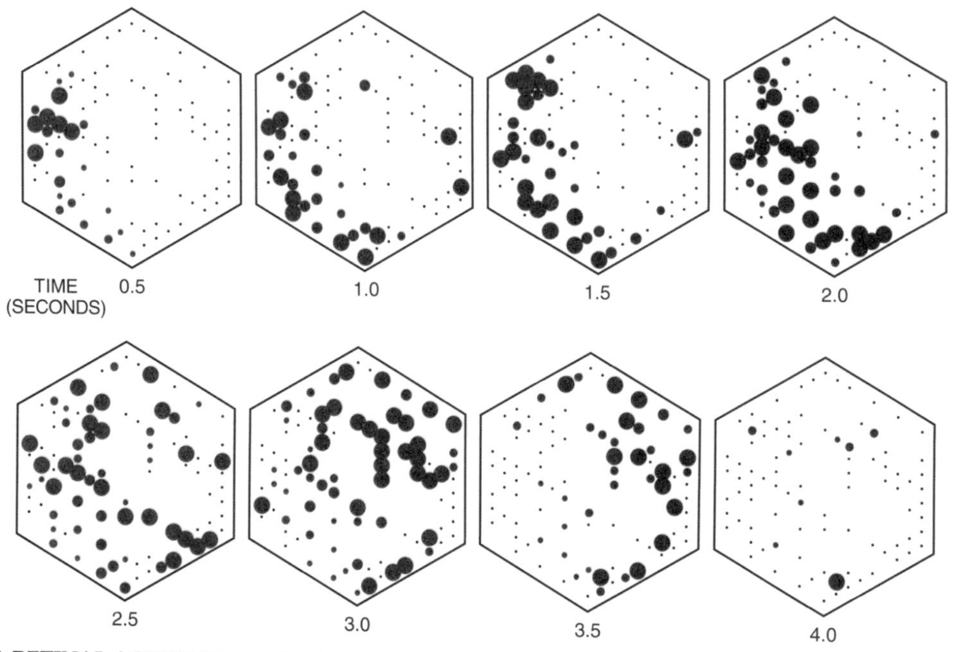

Figure 2.5 RETINAL ACTIVITY, recorded frame by frame every 0.5 second by a hexagonal array of microelectrodes (*black spots*), is locally synchronized. Each diagram represents the pattern and intensity of action-potential firing (*red*) of individual ganglion cells. The wave of retinal activity sweeps across from the lower left to the top right of the retina.

low from Australia visiting my laboratory, succeeded in carefully dissecting the retinas and in tailoring special fluids necessary to maintain the living tissue for hours in a healthy condition.

When neonatal ferret retinas were placed on the multielectrode array, we simultaneously recorded the spontaneously generated action potentials of as many as 100 cells. The work confirmed the in vivo results of Galli-Resta and Maffei. All cells on the array fired within about five seconds of one another, in a predictable and rhythmic pattern. The bursts of action potentials lasted several seconds and were followed by long silent pauses that persisted from 30 seconds to two minutes. This observation showed that the activity of ganglion cells is indeed correlated. Further analysis demonstrated that the activity of neighboring cells is more highly correlated than that of distant cells on the array.

Even more remarkable, the spatial pattern of firing resembled a wave of activity that swept across the retina at about 100 microns per second (about one tenth to one hundredth the speed of an ordinary action potential). After the silent period, another wave was generated but in a completely different and random direction. We found that these spontaneously generated retinal waves are present throughout the period when eye-specific layers take shape. They disappear just before the onset of visual function (see Figure 2.6).

From an engineering standpoint, these waves seem beautifully designed to provide the required correlations in the firing of neighboring ganglion cells. They also ensure a sufficient time delay, so that the synchronized firing of ganglion cells remains local and does not occur across the entire retina. Such a pattern of firing could help refine the topographic map conveyed by ganglion cell axons to each eye-specific layer. Moreover, the fact that wave direction appears to be entirely random implies that ganglion cells in the two eyes are highly unlikely ever to fire synchronously—a requirement for the formation of the layers.

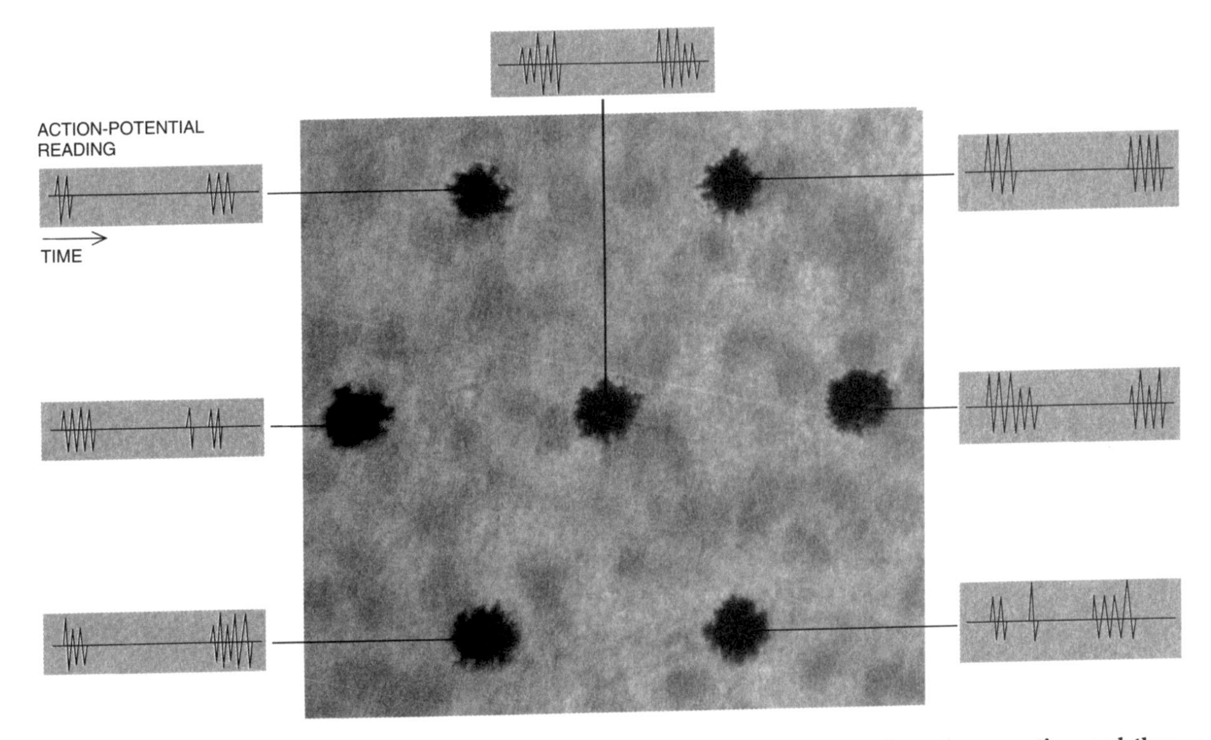

Figure 2.6 ACTION-POTENTIAL READINGS of the developing retina are recorded by microelectrodes (*black spots*). The electrodes detect the small, extracellular currents that flow when the ganglion cells (*stained purple*) fire. All the cells fire at about the same time and then become silent before firing again. The area shown represents about 3 percent of the entire retina.

Future experiments will disrupt the waves in order to determine whether they are truly involved in the development of connections. In addition, it will be important to determine whether the correlations in the firing of neighboring ganglion cells can be detected and used by the cells in the lateral geniculate nucleus to strengthen appropriate synapses and weaken inappropriate ones. This seems likely, since Richard D. Mooney, a postdoctoral fellow in my laboratory, in collaboration with Daniel Madison of Stanford, has shown that long-term potentiation of synaptic transmission between retinal ganglion cell axons and the lateral geniculate nucleus neurons is present during these early periods of development. Thus, at present, we can conclude that even before the onset of function, ganglion cells can spontaneously fire in the correct pattern to fashion the necessary connections.

Is the retina a special case, or might many regions of the nervous system generate their own endogenous activity patterns early in development? Preliminary studies by Michael O'Donovan of the National Institutes of Health suggest that the activity of motor neurons in the spinal cord may also be highly correlated very early in development. It would appear that activity-dependent sorting in this system as well might use spontaneously generated signals. Like those in the visual system, the signals would refine the initially diffuse connections within targets.

The necessity for neuronal activity to complete the development of the brain has distinct advantages. The first is that, within limits, the maturing nervous system can be modified and fine-tuned by experience itself, thereby providing a certain degree of adaptability. In higher vertebrates, this process of refinement can occupy a protracted period. It can begin in utero and, as in the primate visual system, continue well into neonatal life, where it plays an important role in coordinating inputs from the two eyes. The coordination is necessary for binocular vision and stereoscopic depth perception.

Neural activity confers another advantage in development. It is genetically conservative. The alternative—exactly specifying each neural connection using molecular markers—would require an extraordinary number of genes, given the thousands of connections that must be formed in the brain. Using the rules of activity-dependent remodeling described here is far more economical. A major challenge for the future will be to elucidate the cellular and molecular bases for such rules.

The Visual Image in Mind and Brain

In analyzing the distinct attributes of images, the brain invents a visual world. Unusual forms of blindness show what happens when specialized parts of the cortex malfunction.

• • •

Semir Zeki

The study of the visual system is a profoundly philosophical enterprise; it entails an inquiry into how the brain acquires knowledge of the external world, which is no simple matter. The visual stimuli available to the brain do not offer a stable code of information. The wavelengths of light reflected from surfaces change along with alterations in the illumination, yet the brain is able to assign a constant color to them. The retinal image produced by the hand of a gesticulating speaker is never the same from moment to moment, yet the brain must consistently categorize it as a hand. An object's image varies with distance, yet the brain can ascertain its true size (see Figure 3.1).

The brain's task, then, is to extract the constant, invariant features of objects from the perpetually changing flood of information it receives from them. Interpretation is an inextricable part of sensation. To obtain its knowledge of what is visible, the brain cannot therefore merely analyze the images presented to the retina; it must actively construct a visual world. To do so, the brain has developed an elaborate neural mechanism, one so marvelously efficient that it took a century of study before anyone even guessed at its many components. Indeed, when studies of cerebral diseases began to reveal some secrets of the visual brain, neurologists initially dismissed the startling implications as improbable.

The hallmark of that machinery is a complex division of labor. It is manifested anatomically in discrete cortical areas and subregions of areas specialized for particular visual functions; it is manifested pathologically in an inability to acquire knowledge about some aspect of the visual world when the relevant machinery is specifically compromised. Paradoxically, none of this subdivision and specialization within the brain is normally evident at the perceptual level. The visual cortex thus presents us with the intellectual challenge of trying to understand how its components cooperate to give us a unified picture of the world, one that bears no trace of the division of labor within it. There is, to use an old phrase, a great deal more to vision than meets the eye.

This modern conception of the visual brain has evolved only within the past two decades. The early neurologists, starting with those who worked during the late 19th century, saw it very differently. Laboring under the false notion that objects transmitted visual codes in reflected or emitted light, they thought that an image was "impressed" on the retina, much as it would be on a photographic plate. These retinal impressions were subsequently trans-

mitted to the visual cortex, which served to analyze the contained codes. This decoding process led to "seeing." Understanding what was seen—making sense of the received impressions and resolving them into visual objects—was thought to be a separate process that arose through the association of the received impressions with similar ones experienced previously.

This view of how the brain operates, which persisted into the mid-1970s, was therefore also deeply philosophical, although neurologists never acknowledged it as such. It divided sensing from understanding and gave each faculty a separate seat in the cortex. The origin of this dualistic doctrine is obscure, but it bears a resemblance to Immanuel Kant's belief in the two faculties of sensing and of understanding, the former passive and the latter active.

Neurologists saw evidence for their supposition in the fact that the retina connects overwhelmingly to one distinct part of the brain, the striate or primary visual cortex, also known as area V1. This connection is made with high topographic precision: V1 effectively contains a map of the entire retinal field. The retina and V1 are linked through a subcortical structure called the lateral geniculate nucleus, which contains six layers of cells. The four uppermost layers contain cells with small cell bodies and are referred to as the parvocellular layers; the two lowest layers, which have large cell bodies, are the magnocellular layers (see Figure 3.2). Many years ago the late neurologist Salomon E. Henschen of Uppsala University supposed that the function of the large cells was "collecting light" and that of the small cells was registering colors; his basic insight, that the anatomic subdivisions had functional implications, has assumed increasing importance in recent years.

Neurologists at the time found that lesions anywhere along the pathway connecting the retina with V1 created a field of absolute blindness, the extent and position of which corresponded precisely with the size and location of the lesion in V1. That observation led Henschen to conceive of area V1 as the "cortical retina"—the place where "seeing" occurred.

Moreover, the German psychiatrist Paul Emil Flechsig of Leipzig University had shown during the late 19th century that certain regions of the brain, among them V1, had a mature appearance at birth, whereas others, including the cortical regions surrounding V1, continued to develop, as though their maturation depended on the acquisition of experience. For Flechsig and most other neurologists, this observation implied that V1 was "the entering place of the visual radiation into the organ of the psyche," and the areas around it were the repositories of higher "psychic" functions (*Cogitationzentren*) related to sight. Flechsig's theory found support in rather questionable evidence purporting to show that lesions in this so-called visual association cortex, unlike those in V1, might lead to "mind blindness" (*Seelenblindheit*), a condition in which subjects were thought to see but not to comprehend what they saw.

Figure 3.1 WHEN VIEWING AN IMAGE, distinct areas in the cortex analyze it for different visual attributes, such as color, shape and motion. "Seeing" and "understanding" occur simultaneously through the synchronized activities of these cortical areas. The world that one sees is an invention of the visual brain.

Surprisingly, it was research on the visual association cortex that ultimately compromised this dualistic concept of visual brain organization. Work undertaken in the 1970s by John M. Allman and Jon H. Kaas of the University of Wisconsin in the owl monkey and by me in the macaque monkey showed that the visual association cortex—now better referred to as the prestriate cortex—consists of many different cortical areas separated from V1 by another area, V2 (see Figure 3.3). A turning point in our understanding of how the brain constructs the visual image came subsequently, with my demonstration that these areas are individually specialized to undertake different tasks.

In my physiological studies, I presented macaques with a range of stimuli (colors, lines of various orientations and dots moving in different directions) and, using electrodes, monitored the activity of cells in the prestriate cortex. The results showed that all the cells in a prestriate area called V5 are responsive to motion, that most are directionally selective and that none is concerned with the color of the moving stimulus. These facts suggested to me that V5 is specialized for visual motion. (Neuroanatomic terminology is not always uniform; some investigators prefer the label MT to V5.)

In contrast, I found that the overwhelming majority of cells in another area, V4, are to some extent selective for specific wavelengths of light and that many are selective for line orientation, the constitu-

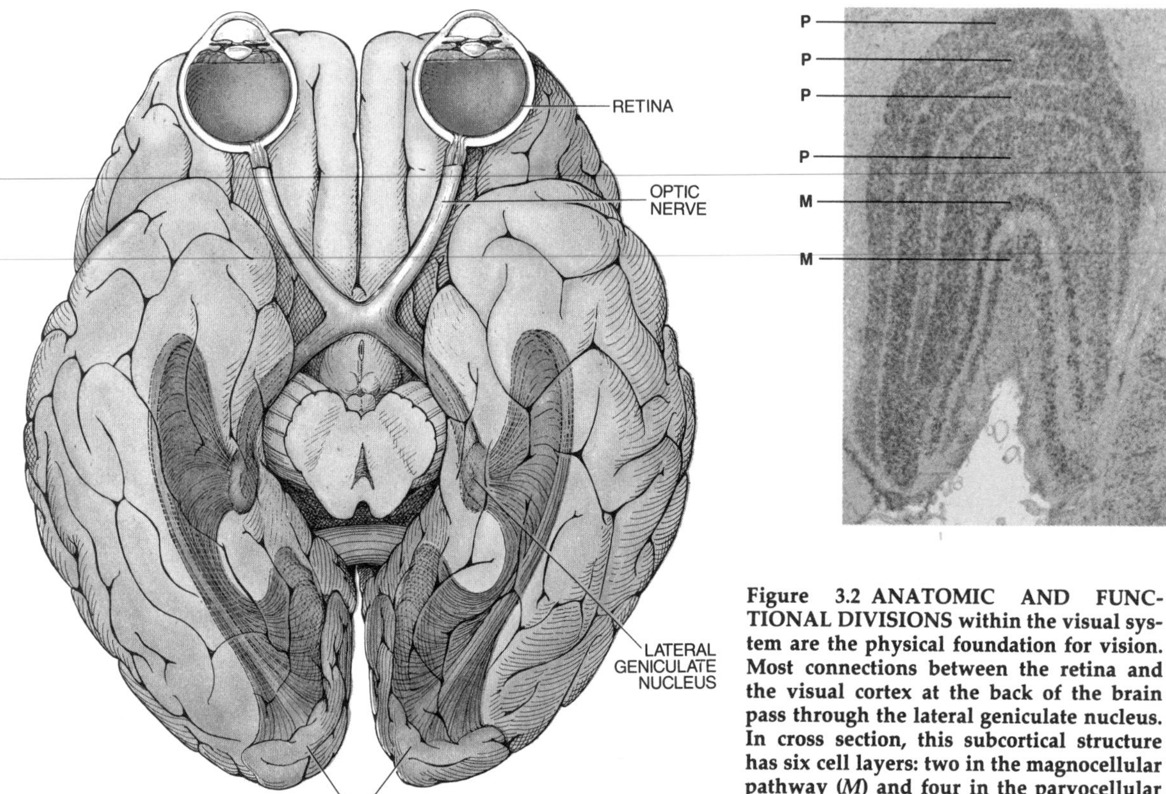

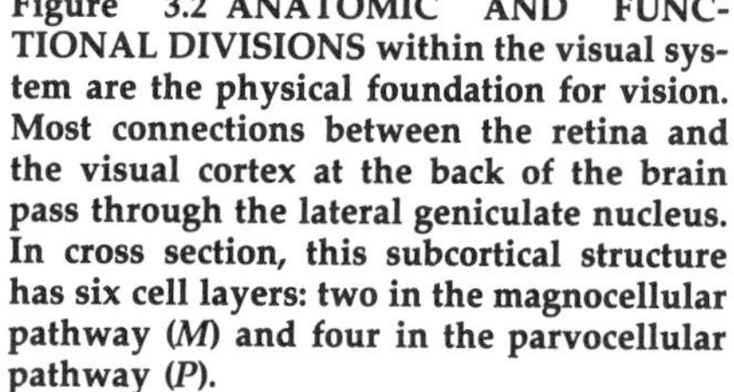

Figure 3.2 ANATOMIC AND FUNCTIONAL DIVISIONS within the visual system are the physical foundation for vision. Most connections between the retina and the visual cortex at the back of the brain pass through the lateral geniculate nucleus. In cross section, this subcortical structure has six cell layers: two in the magnocellular pathway (*M*) and four in the parvocellular pathway (*P*).

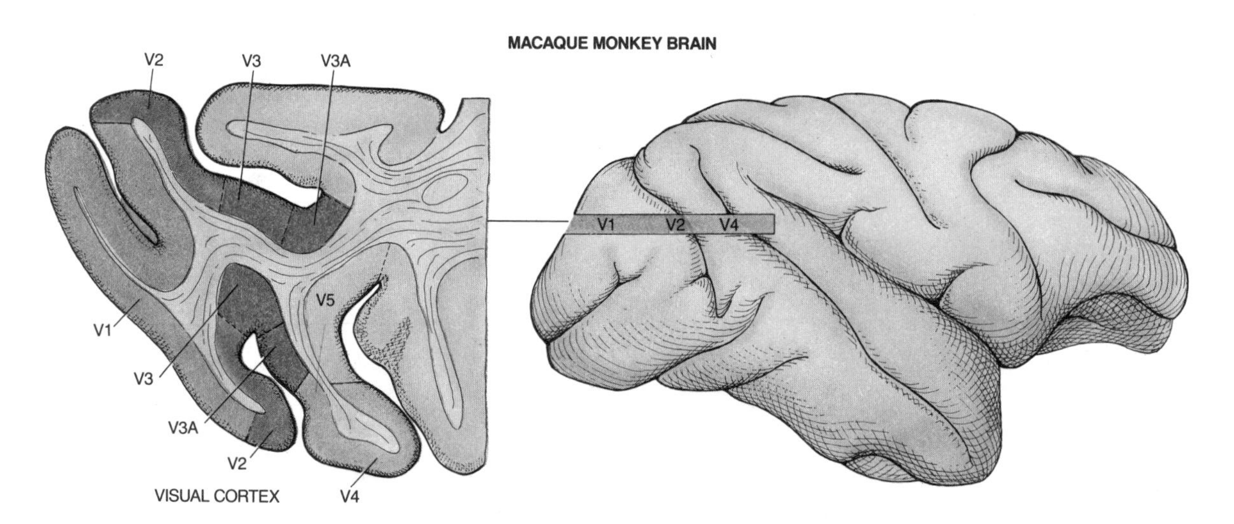

Figure 3.3 VISUAL CORTEX of the macaque monkey has been studied in detail. A cross section through the brain (*left*) at the level indicated (*right*) shows part of the primary visual cortex (V1) and some of the other visual areas in the prestriate cortex (V2–V5).

ents of form, as well. By far most of the cells in two further adjoining areas, V3 and V3A, are also selective for form but like the cells of area V5 are largely indifferent to the color of the stimulus.

These studies led me to propose in the early 1970s the concept of functional specialization in the visual cortex, which supposes that color, form, motion and possibly other attributes of the visible world are processed separately. Because the preponderance of input to the specialized areas comes from V1, a corollary of this finding was that V1 must also show a functional specialization, as must area V2, which receives input from V1 and connects with the same specialized areas. These two areas must, in a sense, act as a kind of post office, parceling out different signals to the appropriate areas.

In recent years, new tissue-staining techniques in combination with physiological studies have provided a startling confirmation of that theory. They have also allowed us to trace these specializations from V1 throughout the prestriate cortex.

With the advent of positron emission tomography (PET), which can measure increases in regional cerebral blood flow when people perform specific tasks, my colleagues at the Hammersmith Hospital in London and I have begun to apply these findings, which were derived from experiments on monkeys, to a direct study of the human brain. We found that when normal-seeing humans view a Land color Mondrian (an abstract painting containing no recognizable objects), the highest increase in regional cerebral blood flow occurs in a structure named the fusiform gyrus. By analogy with a similar region in macaque monkeys, we refer to this cortical area as human V4. The results are very different when subjects view a pattern of moving black-and-white squares: the highest cerebral blood flow then occurs in a more lateral area, quite separate from V4, which we call human V5 (see Figure 3.4).

This demonstration of the separation of motion and color processing constitutes direct evidence that functional specialization is also a feature of the human visual cortex. The PET studies reveal another interesting feature: under both conditions of stimulation, area V1 (and probably the adjoining area V2) also showed marked increases in regional cerebral blood flow. As in the monkeys, these regions, too, must be distributing signals to different areas of the prestriate cortex.

The key to the distribution system in these areas lies in their structural and functional organization. Area V1 is unusually rich in cell layers, yet it reveals an even richer architecture if one examines it with a staining technique first applied by Margaret Wong-Riley of the Medical College of Wisconsin in Milwaukee. The organelles known as mitochondria contain a metabolic enzyme called cytochrome oxidase that makes energy available to a cell. By staining a region of the brain for that enzyme, researchers can identify which cells have the greatest metabolic activity.

When so stained, the metabolic architecture of V1 is characterized by columns of cells that extend from the cortical surface to the underlying nerve tissue called white matter. If viewed in sections cut parallel to the cortical surface, these columns appear as heavily stained blobs or puffs, separated from one another by more lightly stained interblob regions. At Harvard Medical School, Margaret Livingstone and David H. Hubel found that wavelength-selective cells are concentrated in the blobs of V1, whereas form-selective cells are concentrated in the interblobs.

The columns are especially prominent in the second and third layers of V1, which receive input from parvocellular layers of the lateral geniculate nucleus. The cells in those parts of the lateral geniculate nucleus respond in a strong, sustained way to visual stimuli, and many of them are concerned with color.

A distinct set of structures can be seen in layer 4B of V1, which receives input from the magnocellular layers of the lateral geniculate nucleus, whose cells respond transiently to stimuli and are mostly indifferent to color. Layer 4B projects to areas V5 and V3. The cells in layer 4B that connect with V5 are clustered into small patches that are isolated by cells connected to other visual areas. In short, the organization of layer 4B in V1 suggests that certain parts of it are specialized for motion perception and are segregated from regions that handle other attributes.

Like V1, area V2 has a special metabolic architecture. In the case of V2, however, that architecture takes the shape of thick stripes and thin stripes separated from one another by more lightly staining interstripes. As work done by Edgar A. DeYoe and David C. Van Essen of the California Institute of Technology, by Hubel and Livingstone, and by Stewart Shipp of University College, London, and me has shown, cells selective for wavelength congregate in the thin stripes, and cells selective for directional motion are found in the thick stripes.

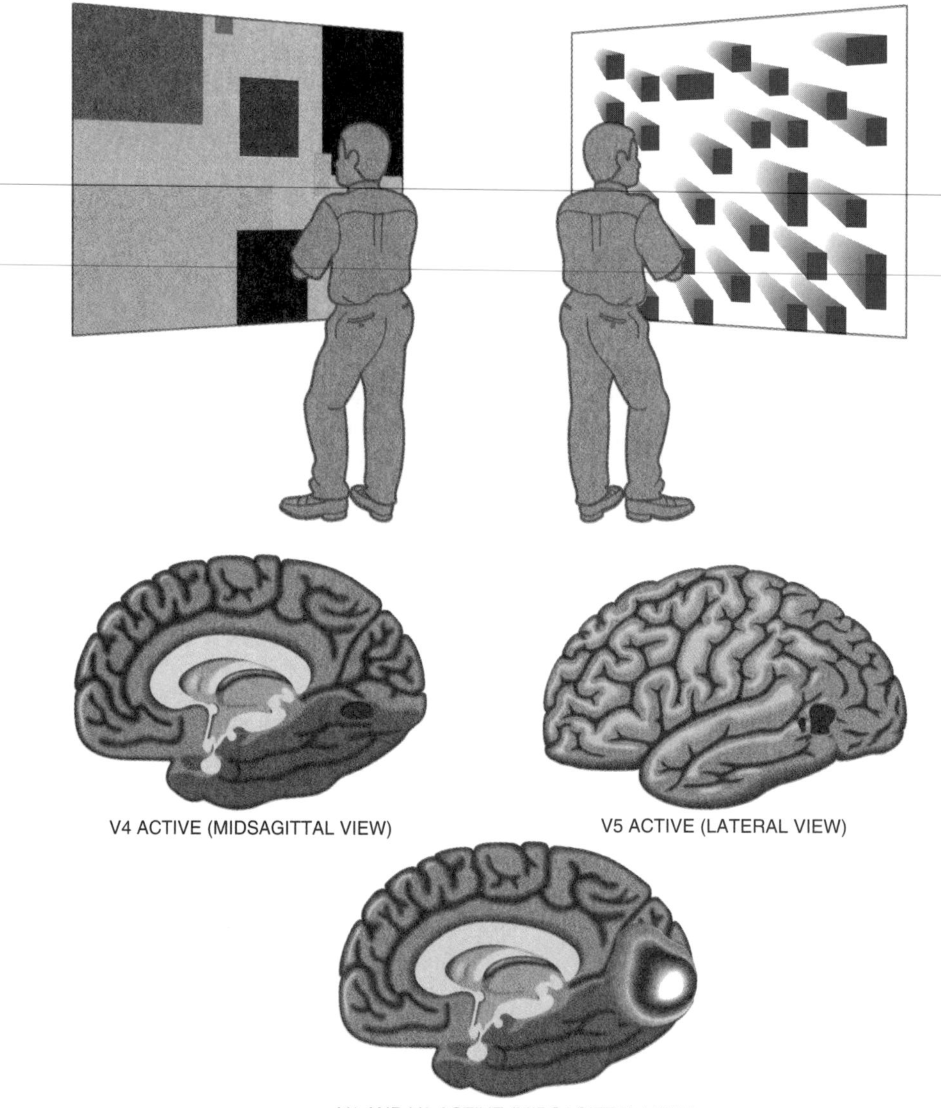

Figure 3.4 DISSIMILAR IMAGES stimulate different regions of the visual cortex. A brightly colored Mondrian causes area V4 to become highly active, as shown by tests of regional cerebral blood flow. Black-and-white moving images trigger activity in area V5. Both types of images lead to activity in areas V1 and V2, which have less specialized functions and distribute signals to other cortical areas.

Cells sensitive to form are distributed in both the thick stripes and the interstripes.

V1 and V2 might therefore be said to contain pigeonholes into which the different signals are assembled before being relayed to the specialized visual areas. The cells in these pigeonholes have small receptive fields; that is, they respond only to stimuli falling on a small region of the retina. They also register information about only a specific attribute of the world within that receptive field. It is as though V1 and V2 were undertaking a piecemeal analysis of the entire field of view.

These facts allow us to delineate four parallel systems concerned with different attributes of vision—one for motion, one for color and two for form (see Figure 3.5). The two that are computationally most distinct from each other are the motion and color systems. For the motion system, the pivotal prestriate area is V5; its inputs run from the retina, through the magnocellular layers of the lateral geniculate nucleus, to layer 4B of V1. From there the signals pass to V5, both directly and through the thick stripes of V2. The color system depends on area V4; its inputs pass through the parvocellular layers of the lateral geniculate nucleus to the blobs of V1, then proceed to V4 directly or through the thin stripes of V2.

Of the two form systems, one is intimately linked to color, and the other is independent of it. The first is based on V4 and derives its inputs from the par-

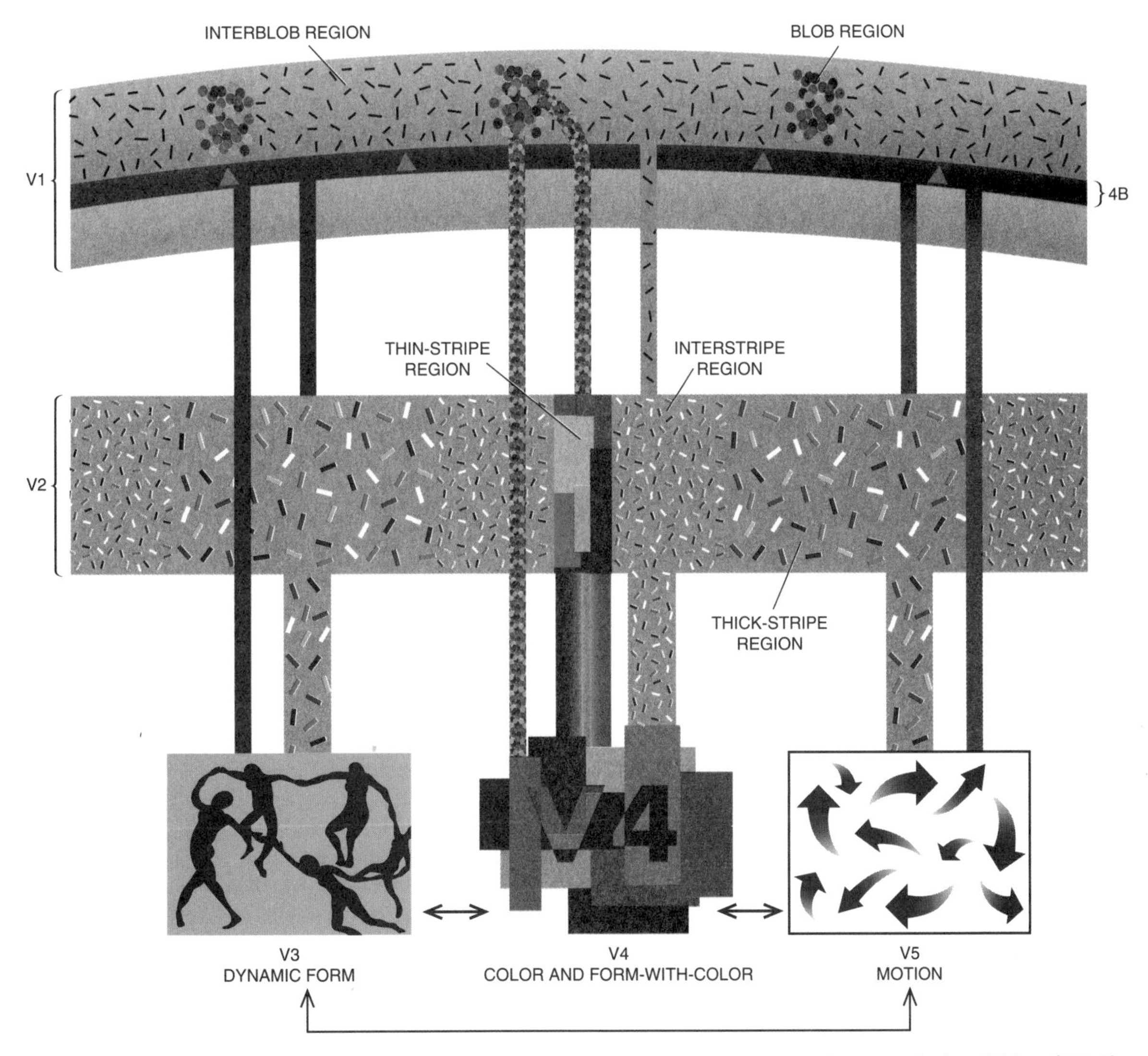

Figure 3.5 FOUR PERCEPTUAL PATHWAYS within the visual cortex have been identified. Color is seen when wavelength-selective cells in the blob regions of V1 send signals to specialized area V4 and also to the thin stripes of V2, which connect with V4. Form in association with color depends on connections between the interblobs of V1, the interstripes of V2 and area V4. Cells in layer 4B of V1 send signals to specialized areas V3 and V5 directly and also through the thick stripes of V2; these connections give rise to the perception of motion and dynamic form.

vocellular layers of the lateral geniculate nucleus by way of the interblobs of V1 and the interstripes of V2. The second is based on V3 and is more concerned with dynamic form—the shapes of objects in motion. It derives its inputs from the magnocellular layers of the lateral geniculate nucleus through layer 4B of V1; the signals then proceed to V3 both directly and through the thick stripes of V2.

Although these four systems are distinct, the anatomy of areas V1 and V2 offers many opportunities for the pigeonholes to communicate with one another, as do the direct connections between the specialized visual areas. Hence, there is an admixture of the parvocellular and magnocellular signals, which the prestriate areas use in different ways to execute their functions.

This remarkable segregation of functions is reflected in some of the pathologies afflicting the visual cortex. Lesions in specific cortical areas produce correspondingly specific visual syndromes that may be far less debilitating than total blindness yet still severe enough to drive patients to distraction and despair. Lesions in area V4 lead to achromatopsia, in which patients see only in shades of gray. This syndrome is different from simple color blindness: not only do such patients fail to see or know the world in color, they cannot even recall colors from a time before the lesion formed. Nevertheless, if their retinas and V1 regions are healthy, their knowledge of form, depth and motion remains intact.

Similarly, a lesion in area V5 produces akinetopsia, in which patients neither see nor understand the world in motion. While at rest, objects may be perfectly visible to them, but motion relative to them causes the objects to vanish. The other attributes of vision remain unscathed—a specificity that results directly from functional differentiation in the human visual cortex.

Given the separation of form and color in the cortex, it is perhaps a little surprising that no one has ever reported a complete and specific loss of form vision. A partial explanation is that such a deficit would require the obliteration of areas V3 and V4 to eliminate both form systems. Area V3 forms a ring around V1 and V2. Consequently, a lesion large enough to destroy all of V3 and V4 would almost certainly destroy V1 as well and thus cause total blindness.

Some patients with lesions in the prestriate cortex do suffer from a degree of form imperception, often coupled with achromatopsia. These people commonly experience far greater difficulty when identifying stationary forms than when the same forms are in motion. They frequently prefer watching television to watching the real world, because television is dominated by moving images. When faced with stationary objects, these patients often resort to the strategy of moving their heads to simplify the task of identification. These observations suggest that they acquire their knowledge of forms through the dynamic form system based in area V3.

The functional specialization in the visual cortex also manifests itself in a syndrome that I have called the chromatopsia ("color vision") of carbon monoxide poisoning. This condition has been described sporadically but not infrequently in the medical literature, although it was never taken seriously until the functional specialization was discovered. Some people who survive the lethal effects of smoke inhalation during fires often suffer diffuse cortical damage from carbon monoxide poisoning, which deprives tissues of oxygen. As a result, these patients often have vision that is severely compromised in all respects except one: their color vision is affected only mildly if at all. Because color is the only kind of visual knowledge available to them, the patients try—often unsuccessfully—to identify all objects solely on the basis of color. They may, for example, misidentify all blue objects as "ocean."

The precise cause of this strange chromatopsia is unknown. The metabolically active blobs of V1 and the thin stripes of V2, both of which are concerned with color, do have unusually high concentrations of blood vessels nourishing them. It is therefore probable that these regions are relatively spared from damage because their rich blood supply renders them less vulnerable to oxygen deprivation.

In summary, then, we know that a total lesion in area V1 produces a complete inability to acquire any visual information and that a lesion in one of the specialized areas makes a corresponding attribute of the visual world inaccessible and incomprehensible. What would happen, we might ask, if signals from the lateral geniculate nucleus were routed directly to the specialized areas, thus bypassing V1 altogether? Nature has actually done that experiment for us, and the resulting phenomenon provides more important insights into the functioning of the visual cortex.

The phenomenon is known as blindsight; it was first described by Ernst Pöppel of the University of

Munich and his colleagues and later studied in great detail by Lawrence Weiskrantz of the University of Oxford and his colleagues. People with this condition are totally blind because of lesions in area V1. Yet if they are forced to guess, they can discriminate correctly among a wide variety of visual stimuli. They can, for example, distinguish between motion in different directions or between different wavelengths of light. Their abilities are imperfect and not completely reliable, but they are better than random guessing. Nevertheless, blindsight patients are not consciously aware of having seen anything at all, and they are often surprised that their "guesses" should have been so accurate.

The basis for this discrimination almost certainly resides in a small but direct connection between the lateral geniculate nucleus and the prestriate cortex, as uncovered by Masao Yukie of the Tokyo Metropolitan Institute for Neurosciences and by Wolfgang Fries of the University of Munich. Alternatively, some other as yet undiscovered subcortical connection to the specialized areas may be responsible. In any event, neurologists have good reason to suppose that in blindsight patients, visual signals reach the prestriate cortex.

Blindsight patients are people who "see" but do not "understand." Because they are unaware of what they have seen, they have not acquired any knowledge. In short, their "vision," which can be elicited only in laboratory situations, is quite useless. Thus, for the visual cortex to do its job of acquiring a knowledge of the world, a healthy V1 area is essential. V1 (and, by extension, V2) may be necessary because it begins to process information for further refinement by the specialized areas or because the results of the processing performed by the specialized areas are referred back to it.

The clinical literature holds many other examples that illuminate how the preprocessing in areas V1 and V2 may contribute directly and explicitly to perception. Damage to V5 can destroy the ability to discriminate the direction or coherence of motions. Yet as Robert F. Hess of the University of Cambridge and his colleagues found, such akinetopsic patients may still be aware that motion of some type is taking place, presumably because of signaling by cells in V1 and V2 (and possibly in other areas that receive magnocellular pathway signals). Similarly, an achromatopsic patient with a V4 lesion, whom I studied with Fries, could discriminate between different wavelengths because of his largely intact V1 even though he could no longer interpret the wavelength information as color (see boxed figure "The World Seen through a Damaged Cortex").

Further insights come from a comparison of the residual form-vision capacities in two other patients who have cortical lesions that are, in a way, complementary. The first patient has a diffuse cortical lesion, caused by carbon monoxide poisoning, that affects area V1. He has terrific difficulty copying even simple forms, such as geometric shapes or letters of the alphabet, because the form-detecting system in his V1 area is so severely compromised.

The second patient has an extensive prestriate lesion from a stroke that has generally spared area V1. He can reproduce a sketch of St. Paul's Cathedral with greater skill than many normal people, although it takes him a great deal of time to do so. Yet this patient has no comprehension of what he has drawn. Because his V1 system is largely intact, he can identify the local elements of form, such as angles and simple shapes, and accurately copy the lines he sees and understands. The prestriate lesion, however, prevents him from integrating the lines into a complex whole and recognizing it as a building. The patient sees and understands only what the limited capacity of his intact system allows.

The residual capacity in such patients unmasks an important feature of the organization of the visual cortex, namely, that none of the visual areas—not even "post office" areas V1 and V2—serves merely to relay signals to other areas. Instead each is part of the machinery that actively transforms the incoming signals and may contribute explicitly, if incompletely, to perception.

The profound division of labor within the visual cortex naturally raises the question of how the specialized areas interact to provide a unified image. The simplest way would be for all the specialized areas to communicate the results of their operations to one master area, which would then synthesize the incoming information. Philosophically, that solution begs the question, because one must then ask who or what looks at the composite image, and how it does so. That problem is beside the point, however, because the anatomic evidence shows no single master area to which all the antecedent areas exclusively connect. Instead the specialized areas connect with one another, either directly or through other areas.

For example, areas V4 and V5 connect directly and reciprocally with each other. Both of them also project to the parietal and temporal regions of the

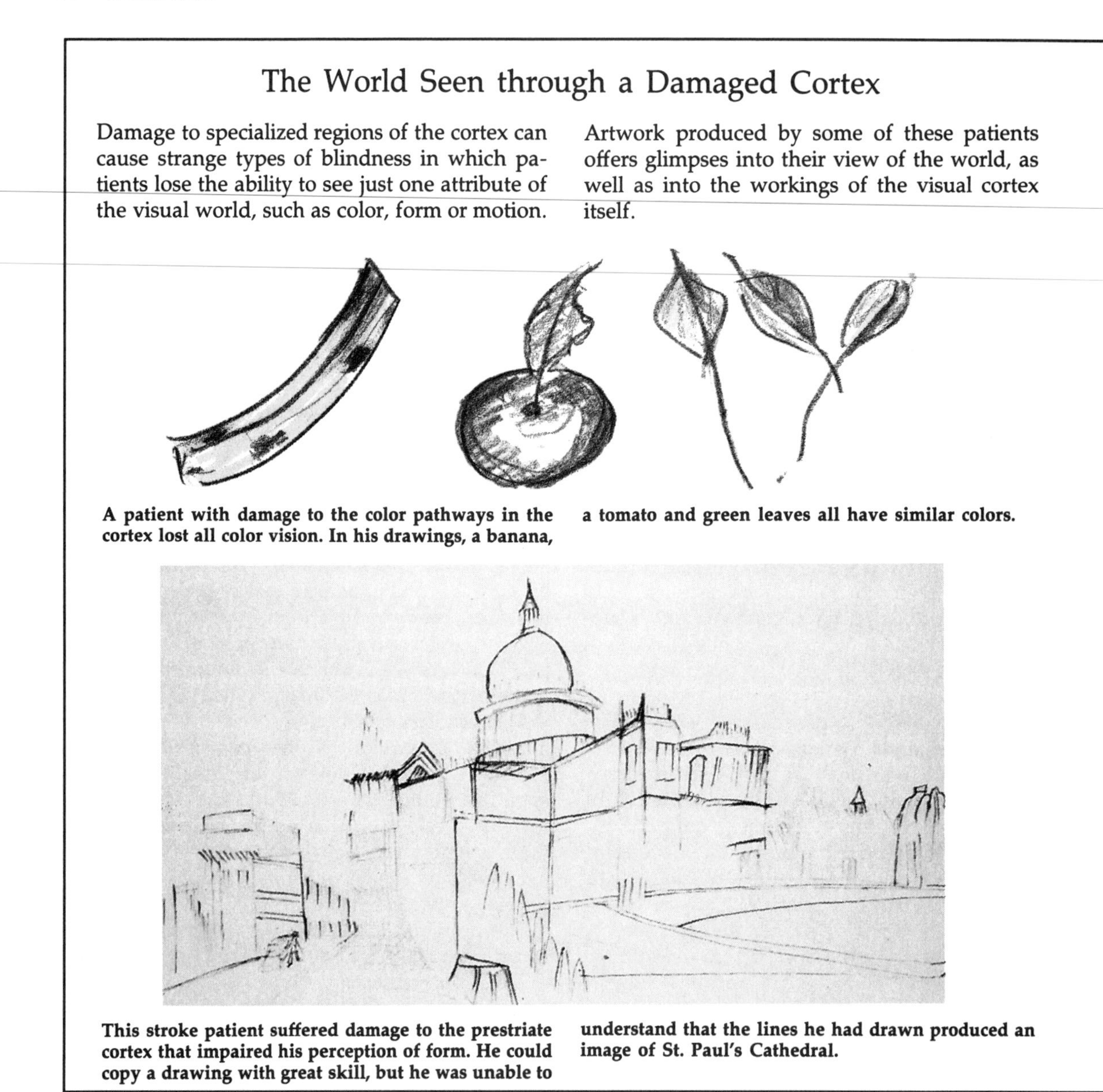

The World Seen through a Damaged Cortex

Damage to specialized regions of the cortex can cause strange types of blindness in which patients lose the ability to see just one attribute of the visual world, such as color, form or motion. Artwork produced by some of these patients offers glimpses into their view of the world, as well as into the workings of the visual cortex itself.

A patient with damage to the color pathways in the cortex lost all color vision. In his drawings, a banana, a tomato and green leaves all have similar colors.

This stroke patient suffered damage to the prestriate cortex that impaired his perception of form. He could copy a drawing with great skill, but he was unable to understand that the lines he had drawn produced an image of St. Paul's Cathedral.

brain, but as my work with my colleagues has shown, the outputs from each area occupy their own unique territory within the receiving region. Direct overlap between the signals from V4 and V5 is minimal. It is as if the cortex wishes to maintain the separation of the distinct visual signals—a strategy it also employs in memory and other systems (see Chapter 6, "Working Memory and the Mind," by Patricia S. Goldman-Rakic). Any integration of the signals within the parietal or temporal regions must occur through local "wiring" that connects the inputs.

In fact, integration of the visual information is a monumental task that necessitates a vast network of

When an achromatopsic (color-blind) patient was shown a Land color Mondrian (*left*) and asked to reproduce it, he was able to copy the shapes in the painting successfully. The colors within the blocks eluded him (*right*).

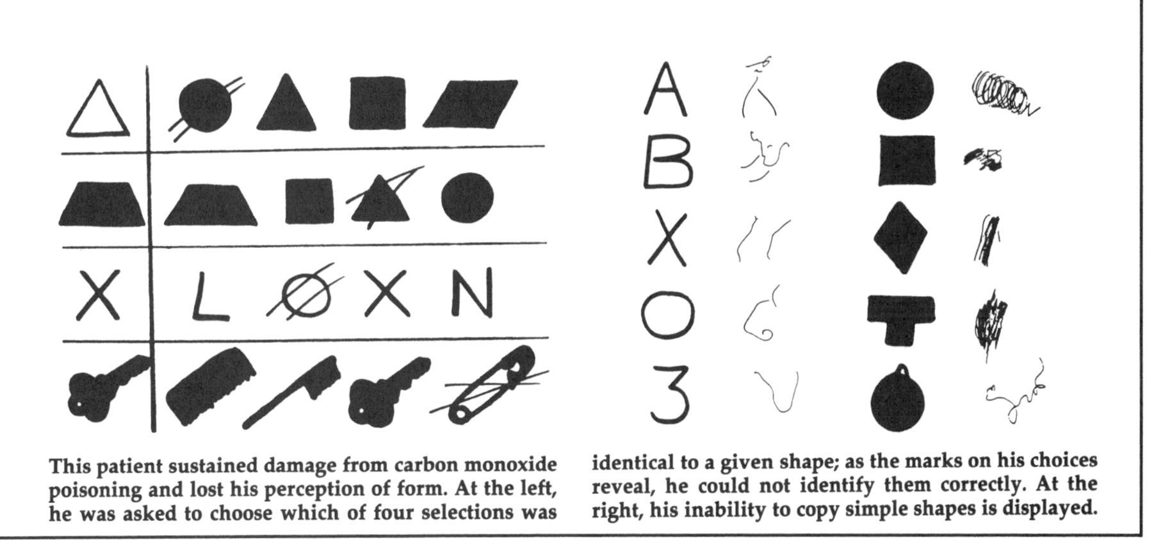

This patient sustained damage from carbon monoxide poisoning and lost his perception of form. At the left, he was asked to choose which of four selections was identical to a given shape; as the marks on his choices reveal, he could not identify them correctly. At the right, his inability to copy simple shapes is displayed.

anatomic links between the four parallel systems at every level, because each level contributes explicitly to perception. Integration also creates some formidable problems. To understand coherent motion, for example, the brain must determine which features in the field of view are moving in the same direction and at the same speed. The motion-sensing cells in the specialized areas are able to make those comparisons because they have larger receptive fields than do their antecedent counterparts in V1.

Yet because their receptive fields are larger, these cells are inherently less efficient at pinpointing the position of any one stimulus within the visual field.

For the brain to make spatial sense of the integrated information, the information must somehow be referred to an area that has a more precise topographic map of the retina and hence of the visual field. Of all the visual areas, the one with the most precise map is area V1, followed by V2. The specialized areas must therefore send information back to V1 and V2 so that the results of the comparisons can be mapped back onto the visual field.

Reentrant connections, which allow information to flow both ways between different areas, are also essential for resolving conflicts between cells that have different capabilities and are responding to the same stimulus. A good example of such conflicts is found in the responses of cells in V1 and V2 to illusory contours, such as those of the Kanizsa triangle (see Figure 3.6).

In this famous illusion, a normal observer perceives a triangle among the presented shapes even though the lines forming the triangle are incomplete; the brain creates lines where there are none. As Rudiger von der Heydt and Esther Peterhans of the Zurich University Hospital have demonstrated, the form-selective cells in V1 do not respond to the illusion and do not signal that a line is present. The V2 cells receive their inputs from V1, but because the V2 cells have larger receptive fields and more analytic functions, they do respond to the illusion by "inferring" the presence of a line. To settle the conflict, the V2 cells must have reentrant inputs to their counterparts in V1.

Another difficulty that arises from the process of integration is the binding problem. Cells responding to the same object in the field of view may be scattered throughout V1. Something must therefore bind together the signals from those cells so that they are treated as belonging to the same object and not separate ones. The problem becomes even more thorny when cells in two or more visual areas respond to different attributes of the same object.

One way to resolve the problem is for the cells to fire in temporal synchrony. In practice, such synchrony does occur, at least to some extent, among cells that are anatomically connected to one another, as work by Wolf J. Singer of the Max Planck Institute for Brain Research in Frankfurt and his colleagues has shown. We must then face the problem, however, of who or what determines that the firing should be synchronous. Reentrant inputs provide at least a partial solution by linking the output of one area to other areas that sent it information.

These problems have led me and my colleagues to develop a theory of multistage integration. It hypothesizes that integration does not occur in a single step through a convergence of output onto a master area, nor does integration have to be postponed until all the visual areas have completed their individual operations. Instead the integration of visual information is a process in which perception and comprehension of the visual world occur simultaneously.

The anatomic requirement for multistage integration is immense, because it involves reentrant connections between all the specialized areas as well as with areas V1 and V2, which feed them signals. Our studies indicate that such a network of reentrant connections does exist.

The reentrant inputs to areas V1 and V2 differ fundamentally from their forward connections to

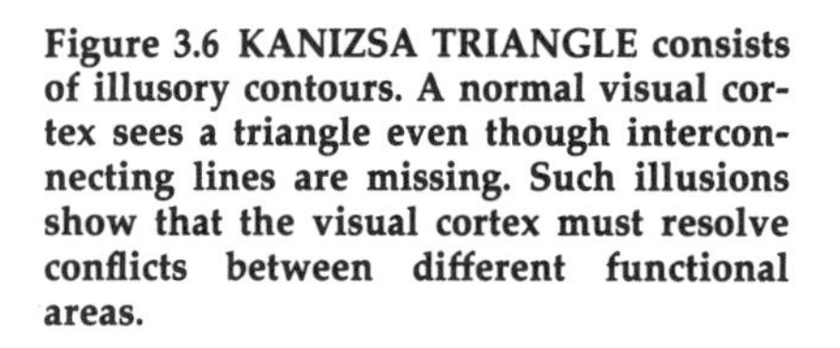
Figure 3.6 KANIZSA TRIANGLE consists of illusory contours. A normal visual cortex sees a triangle even though interconnecting lines are missing. Such illusions show that the visual cortex must resolve conflicts between different functional areas.

the specialized areas. The forward projections are patchy and discrete, because segregated groups of cells in V1 and V2 send their outputs to specialized areas with corresponding visual attributes. The return projections, however, are diffuse and fairly nonspecific. For example, whereas V5 receives input only from select groups of cells in layer 4B of V1, the return input from V5 to layer 4B is diffuse and encompasses the territory of all the cells in the layer, including ones that project into V3. This reentrant system can thus serve three purposes simultaneously: it can unite and synchronize the signals for form and motion found in two different visual pathways; it can refer information about motion back to an area with an accurate topographic map; it can integrate motion information from V5 with form information on its way to V3.

Similarly, whereas the output from V2 to the specialized areas is highly segregated, the return input from those areas to V2 is diffuse. V4 projects back not only to the thin stripes and the interstripes, from which it receives its input, but also to the thick stripes, from which it does not. This reentrant system can therefore help unite signals dealing with form, motion and color.

It is becoming increasingly evident that the entire network of connections within the visual cortex, including the reentrant connections to V1 and V2, must function healthily for the brain to gain complete knowledge of the external world. Yet as patients with blindsight have shown, knowledge cannot be acquired without consciousness, which seems to be a crucial feature of a properly functioning visual apparatus. Consequently, no one will be able to understand the visual brain in any profound sense without tackling the problem of consciousness as well (see Chapter 11, "The Problem of Consciousness," by Francis Crick and Christof Koch).

The past two decades have brought neurologists many marvelous discoveries about the visual brain. Moreover, they have led to a powerful conceptual change in our view of what the visual brain does and how it accomplishes its functions. It is no longer possible to divide the process of seeing from that of understanding, as neurologists once imagined, nor is it possible to separate the acquisition of visual knowledge from consciousness. Indeed, consciousness is a property of the complex neural apparatus that the brain has developed to acquire knowledge.

Thus, our inquiry into the visual brain takes us into the very heart of humanity's inquiry into its own nature. This is not to say that understanding the workings of the visual brain will resolve the problem of consciousness—far from it. But it is a good beginning.

The Biological Basis of Learning and Individuality

Recent discoveries suggest that learning engages a simple set of rules that modify the strength of connections between neurons in the brain. These changes play an important role in making each individual unique.

• • •

Eric R. Kandel and Robert D. Hawkins

Over the past several decades, there has been a gradual merger of two originally separate fields of science: neurobiology, the science of the brain, and cognitive psychology, the science of the mind. Recently the pace of unification has quickened, with the result that a new intellectual framework has emerged for examining perception, language, memory and conscious awareness. This new framework is based on the ability to study the biological substrates of these mental functions. A particularly fascinating example can be seen in the study of learning. Elementary aspects of the neuronal mechanisms important for several different types of learning can now be studied on the cellular and even on the molecular level. The analysis of learning may therefore provide the first insights into the molecular mechanisms underlying a mental process and so begin to build a bridge between cognitive psychology and molecular biology.

Learning is the process by which we acquire new knowledge, and memory is the process by which we retain that knowledge over time. Most of what we know about the world and its civilizations we have learned. Thus, learning and memory are central to our sense of individuality. Indeed, learning goes beyond the individual to the transmission of culture from generation to generation. Learning is a major vehicle for behavioral adaptation and a powerful force for social progress. Conversely, loss of memory leads to loss of contact with one's immediate self, with one's life history and with other human beings.

Until the middle of the 20th century, most students of behavior did not believe that memory was a distinct mental function independent of movement, perception, attention and language. Long after those functions had been localized to different

Figure 4.1 MIRROR DRAWING EXPERIMENT in patients with temporal lobe lesions gave the first hint, in 1960, that there are two distinct types of learning systems. One form, which is spared by the lesions, involves tasks that have an automatic quality such as the skilled movements illustrated in this experiment. The subject, who can see his hand only in the mirror, tries to trace the shape of a star. The second type of learning depends on conscious awareness and cognitive processes and is abolished by the lesions.

MONGOL 482

regions of the brain, researchers still doubted that memory could ever be assigned to a specific region. The first person to do so was Wilder G. Penfield, a neurosurgeon at the Montreal Neurological Institute.

In the 1940s Penfield began to use electrical stimulation to map motor, sensory and language functions in the cortex of patients undergoing neurosurgery for the relief of epilepsy. Because the brain itself does not have pain receptors, brain surgery can be carried out under local anesthesia in fully conscious patients, who can describe what they experience in response to electric stimuli applied to different cortical areas. Penfield explored the cortical surface in more than 1,000 patients. Occasionally he found that electrical stimulation produced an experiential response, or flashback, in which the patients described a coherent recollection of an earlier experience. These memorylike responses were invariably elicited from the temporal lobes.

Additional evidence for the role of the temporal lobe in memory came in the 1950s from the study of a few patients who underwent bilateral removal of the hippocampus and neighboring regions in the temporal lobe as treatment for epilepsy. In the first and best-studied case, Brenda Milner of the Montreal Neurological Institute described a 27-year-old assembly-line worker, H.M., who had suffered from untreatable and debilitating temporal lobe seizures for more than 10 years. The surgeon William B. Scoville removed the medial portion of the temporal lobes on both sides of H.M.'s brain. The seizure disturbance was much improved. But immediately after the operation, H.M. experienced a devastating memory deficit: he had lost the capacity to form new long-term memories.

Despite his difficulty with the formation of new memories, H.M. still retained his previously acquired long-term memory store. He remembered his name, retained a perfectly good use of language and kept his normal vocabulary; his IQ remained in the range of bright-normal. He remembered well the events that preceded the surgery, such as the job he had held, and he remembered vividly the events of his childhood. Moreover, H.M. still had a completely intact short-term memory. What H.M. lacked, and lacked profoundly, was the ability to translate what he learned from short-term to long-term memory. For example, he could converse normally with the hospital staff, but he did not remember them even though he saw them every day.

The memory deficit following bilateral temporal lobe lesions was originally thought to apply equally to all forms of new learning. But Milner soon discovered that this is not the case. Even though patients with such lesions have profound deficits, they can accomplish certain types of learning tasks as well as normal subjects can and retain the memory of these tasks for long periods. Milner first demonstrated this residual memory capability in H.M. with the discovery that he could learn new motor skills normally (see Figure 4.1). She, and subsequently Elizabeth K. Warrington of the National Hospital for Nervous Diseases in London and Lawrence Weiskrantz of the University of Oxford, found that patients such as H.M. can also acquire and retain memory for elementary kinds of learning that involve changing the strength of reflex responses, such as habituation, sensitization and classical conditioning.

It immediately became apparent to students of behavior that the difference between types of learning that emerged from studies of patients with temporal lobe lesions represented a fundamental psychological distinction—a division in the way all of us acquire knowledge. Although it is still not clear how many distinct memory systems there are, researchers agree that lesions of the temporal lobes severely impair forms of learning and memory that require a conscious record. In accordance with the suggestion of Neal J. Cohen of the University of Illinois and Larry R. Squire of the University of California at San Diego and of Daniel L. Schacter of the University of Toronto, these types of learning are commonly called declarative or explicit. Those forms of learning that do not utilize conscious participation remain surprisingly intact in patients with temporal lobe lesions; they are referred to as nondeclarative or implicit.

Explicit learning is fast and may take place after only one training trial. It often involves association of simultaneous stimuli and permits storage of information about a single event that happens in a particular time and place; it therefore affords a sense of familiarity about previous events. In contrast, implicit learning is slow and accumulates through repetition over many trials. It often involves association of sequential stimuli and permits storage of information about predictive relations between events. Implicit learning is expressed primarily by improved performance on certain tasks without the subject being able to describe just what has been learned, and it involves memory systems that do not draw on the contents of the general knowledge of the individual. When a subject such as H.M. is asked why he performs a given task better after five days of practice than on the first day, he may respond, "What are you talking about? I've never done this task before."

Whereas explicit memory requires structures in the temporal lobe of vertebrates, implicit memory is thought to be expressed through activation of the particular sensory and motor systems engaged by the learning task; it is acquired and retained by the plasticity inherent in these neuronal systems. As a result, implicit memory can be studied in various reflex systems in either vertebrates or invertebrates. Indeed, even simple invertebrate animals show excellent reflexive learning.

The existence of two distinct forms of learning has caused the reductionists among neurobiologists to ask whether there is a representation on the cellular level for each of these two types of learning process. Both the neural systems that mediate explicit memory and those that mediate implicit memory can store information about the association of stimuli. But does the same set of cellular learning rules guide the two memory systems as they store associations, or do separate sets of rules govern each system?

An assumption underlying early studies of the neural basis of memory systems was that the storage of associative memory, both implicit and explicit, required a fairly complex neural circuit. One of the first to challenge this view was the Canadian psychologist Donald O. Hebb, a teacher of Milner. Hebb boldly suggested that associative learning could be produced by a simple cellular mechanism. He proposed that associations could be formed by coincident neural activity: "When an axon of cell A . . . excite[s] cell B and repeatedly or persistently takes part in firing it, some growth process or metabolic change takes place in one or both cells such that A's efficacy, as one of the cells firing B, is increased." According to Hebb's learning rule, coincident activity in the presynaptic and postsynaptic neurons is critical for strengthening the connection between them, a so-called pre-post associative mechanism (see Figure 4.2).

Ladislav Tauc and one of us (Kandel) proposed a second associative learning rule in 1963 while working at the Institute Marey in Paris on the nervous system of the marine snail *Aplysia*. They found that the synaptic connection between two neurons could be strengthened without activity of the postsynaptic cell when a third neuron acts on the presynaptic neuron. The third neuron, called a modulatory neuron, enhances transmitter release from the terminals of the presynaptic neuron. They suggested that this mechanism could take on associa-

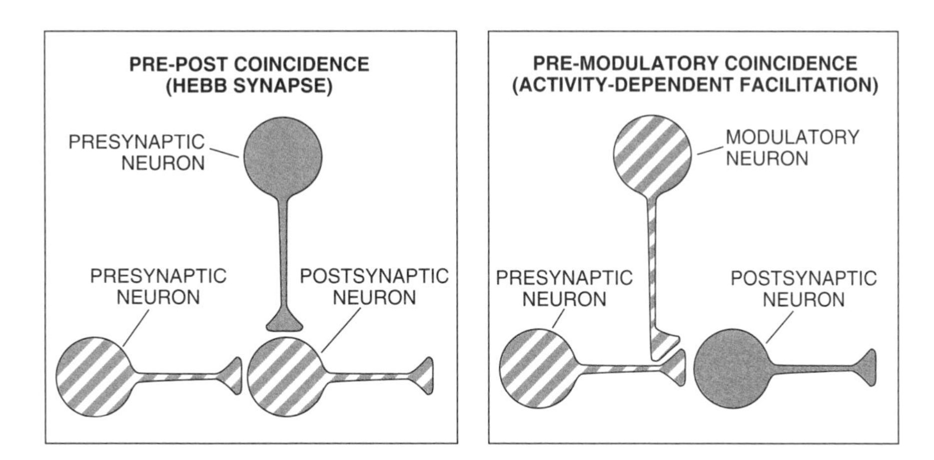

Figure 4.2 TWO CELLULAR MECHANISMS are hypothesized for associative changes in synaptic strength during learning. The pre-post coincidence mechanism, proposed by Donald O. Hebb in 1949, posits that coincident activity in the presynaptic and postsynaptic neurons is critical for strengthening the connections between them. The premodulatory coincidence mechanism proposed in 1963, based on studies in *Aplysia*, holds that the connection can be strengthened without activity of the postsynaptic cell when a third neuron, the modulatory neuron, is active at the same time as the presynaptic neuron. Strips denote neurons in which coincident activity must occur to produce the associative change.

tive properties if the electrical impulses known as action potentials in the presynaptic cell were coincident with action potentials in the modulatory neuron (a pre-modulatory associative mechanism).

Subsequently, we and our colleagues Thomas J. Carew and Thomas W. Abrams of Columbia University and Edgar T. Walters and John H. Byrne of the University of Texas Health Science Center found experimental confirmation. We observed the pre-modulatory associative mechanism in *Aplysia*, where it contributes to classical conditioning, an implicit form of learning. Then, in 1986, Holger J. A. Wigström and Bengt E. W. Gustafsson, working at the University of Göteborg, found that the pre-post associative mechanism occurs in the hippocampus, where it is utilized in types of synaptic change that are important for spatial learning, an explicit form of learning.

The finding of two distinct cellular learning rules, each with associative properties, suggested that the associative mechanisms for implicit and explicit learning need not require complex neural networks. Rather the ability to detect associations may simply reflect the intrinsic capability of certain cellular interactions. Moreover, these findings raised an intriguing question: Are these apparently different mechanisms in any way related? Before considering their possible interrelation, we shall first describe the two learning mechanisms, beginning with the pre-modulatory mechanism contributing to classical conditioning in *Aplysia*.

Classical conditioning was first described at the turn of the century by the Russian physiologist Ivan Pavlov, who immediately appreciated that conditioning represents the simplest example of learning to associate two events. In classical conditioning, an ineffective stimulus called the conditioned stimulus (or more correctly, the to-be-conditioned stimulus) is repeatedly paired with a highly effective stimulus called the unconditioned stimulus. The conditioned stimulus initially produces only a small response or no response at all; the unconditioned stimulus elicits a powerful response without requiring prior conditioning.

As a result of conditioning (or learning), the conditioned stimulus becomes capable of producing either a larger response or a completely new response. For example, the sound of a bell (the conditioned stimulus) becomes effective in eliciting a behavioral response such as lifting a leg only after that sound has been paired with a shock to the leg (the unconditioned stimulus) that invariably produces a leg-lifting response. For conditioning to occur, the conditioned stimulus generally must be correlated with the unconditioned stimulus and precede it by a certain critical period. The animal is therefore thought to learn predictive relations between the two stimuli.

Because *Aplysia* has a nervous system containing only about 20,000 central nerve cells, aspects of classical conditioning can be examined at the cellular level. *Aplysia* has a number of simple reflexes, of which the gill-withdrawal reflex has been particularly well studied. The animal normally withdraws the gill, its respiratory organ, when a stimulus is applied to another part of its body such as the mantle shelf or the fleshy extension called the siphon. Both the mantle shelf and the siphon are innervated by their own populations of sensory neurons. Each of these populations makes direct contact with motor neurons for the gill as well as with various classes of excitatory and inhibitory interneurons that synapse on the motor neurons. We and our colleagues Carew and Walters found that even this simple reflex can be conditioned.

A weak tactile stimulus to one pathway, for example, the siphon, can be paired with an unconditioned stimulus (a strong shock) to the tail. The other pathway, the mantle shelf, can then be used as a control pathway. The control pathway is stimulated the same number of times, but the stimulus is not paired (associated) with the tail shock. After five pairing trials, the response to stimulation of the siphon (the paired pathway) is greater than that of the mantle (the unpaired pathway). If the procedure is reversed and the mantle shelf is paired rather than the siphon, the response to the mantle shelf will be greater than that to the siphon. This differential conditioning is remarkably similar in several respects to that seen in vertebrates.

To discover how this conditioning works, we focused on one component: the connections between the sensory neurons and their target cells, the interneurons and motor neurons. Stimulating the sensory neurons from either the siphon or the mantle shelf generates excitatory synaptic potentials in the interneurons and motor cells. These synaptic potentials cause the motor cells to discharge, leading to a brisk reflex withdrawal of the gill. The unconditioned reinforcing stimulus to the tail activates many cell groups, some of which also cause movement of the gill. Among them are at least three groups of modulatory neurons, in one of which the chemical serotonin is the transmitter. (Neurotransmitters such as serotonin that carry messages between cells are called first messengers; other chemicals known as second messengers relay information within the cell.)

These modulatory neurons act on the sensory neurons from both the siphon and the mantle shelf, where they produce presynaptic facilitation, that is, they enhance transmitter release from the terminals of the sensory neurons. Presynaptic facilitation contributes to a nonassociative form of learning called sensitization, in which an animal learns to enhance a variety of defensive reflex responses after receiving a noxious stimulus [see "Small Systems of Neurons," by Eric R. Kandel; SCIENTIFIC AMERICAN, September 1979]. This type of learning is referred to as nonassociative because it does not depend on pairing between stimuli.

The finding that modulatory neurons act on both sets of sensory neurons—those from the siphon as well as those from the mantle—posed an interesting question: How is the specific associative strengthening of classical conditioning achieved? Timing turned out to be an important element here. For classical conditioning to occur, the conditioned stimulus generally must precede the unconditioned stimulus by a critical and often narrow interval. For conditioning gill withdrawal by tail shock, the interval is approximately 0.5 second. If the separation is lengthened, shortened or reversed, conditioning is drastically reduced or does not occur.

In the gill-withdrawal reflex, the specificity in timing results in part from a convergence of the conditioned and unconditioned stimuli within individual sensory neurons. The unconditioned stimulus is represented in the sensory neurons by the action of the modulatory neurons, in particular the cells in which serotonin is the transmitter. The conditioned stimulus is represented by activity within the sensory neurons themselves. We found that the modulatory neurons activated by the unconditioned stimulus to the tail produce greater presynaptic facilitation of the sensory neurons if the sensory neurons had just fired action potentials in response to the conditioned stimulus. Action potentials in the sensory neurons that occur just after the tail shock have no effect.

This novel property of presynaptic facilitation is called activity dependence. Activity-dependent facilitation requires the same timing on the cellular level as does conditioning on the behavioral level and may account for such conditioning. These results suggest that a cellular mechanism of classical conditioning of the withdrawal reflex is an elaboration of presynaptic facilitation, a mechanism used for sensitization of the reflex. These experiments provided an initial suggestion that there might be a cellular alphabet for learning whereby the mechanisms of more complex types of learning may be elaborations or combinations of the mechanisms of simpler types of learning.

The next piece in the puzzle of how classical conditioning occurs was to discover why the firing of action potentials in the sensory neurons just before the unconditioned tail stimulus would enhance presynaptic facilitation. We had previously found that when serotonin is released by the modulatory neurons in response to tail shock, it initiates a series of biochemical changes in the sensory neurons (see boxed figure "Classical Conditioning in *Aplysia*"). Serotonin binds to a receptor that activates an enzyme called adenylyl cyclase. This enzyme in turn converts ATP, one of the molecules that provides the energy needed to power the various activities of the cell, into cyclic AMP. Cyclic AMP then acts as a second messenger (serotonin is the first messenger) inside the cell to activate another enzyme, a protein kinase. Kinases are proteins that phosphorylate (add a phosphate group to) other proteins, thereby increasing the activity of some and decreasing the activity of others.

The activation of the protein kinase in sensory neurons has several important short-term consequences. The protein kinase phosphorylates potassium channel proteins. Phosphorylation of these channels (or of proteins that act on these channels) reduces a component of the potassium current that normally repolarizes the action potential. Reduction of potassium current prolongs the action potential and thereby allows calcium channels to be activated for longer periods, permitting more calcium to enter the presynaptic terminal. Calcium has several actions within the cell, one of which is the release of transmitter vesicles from the terminal. When, as a result of an increase in the duration of the action potentials, more calcium enters the terminal, more transmitter is released. Second, as a result of protein kinase activity, serotonin acts to mobilize transmitter vesicles from a storage pool to the release sites at the membrane; this facilitates the release of transmitter independent of an increase in calcium influx. In this action, cyclic AMP acts in parallel with another second messenger, protein kinase C, which is also activated by serotonin.

Why should the firing of action potentials in the sensory neurons just before the unconditioned stim-

Classical Conditioning in *Aplysia*

The diagrams (*left*) trace one of the pathways involved in classical conditioning of the gill-withdrawal reflex in *Aplysia*. An increase in the release of neurotransmitter due to activity-dependent facilitation is a mechanism that contributes to conditioning. In activity-dependent facilitation (*right*) serotonin released from the modulatory neuron by the unconditioned stimulus activates adenylyl cyclase in the sensory neuron. When the sensory neuron is active, calcium levels are elevated within the cell. The calcium binds to calmodulin, which binds to adenylyl cyclase, enhancing its ability to synthesize cyclic AMP. The cyclic AMP activates protein kinase, which leads to the release of a greater amount of transmitter than would occur normally.

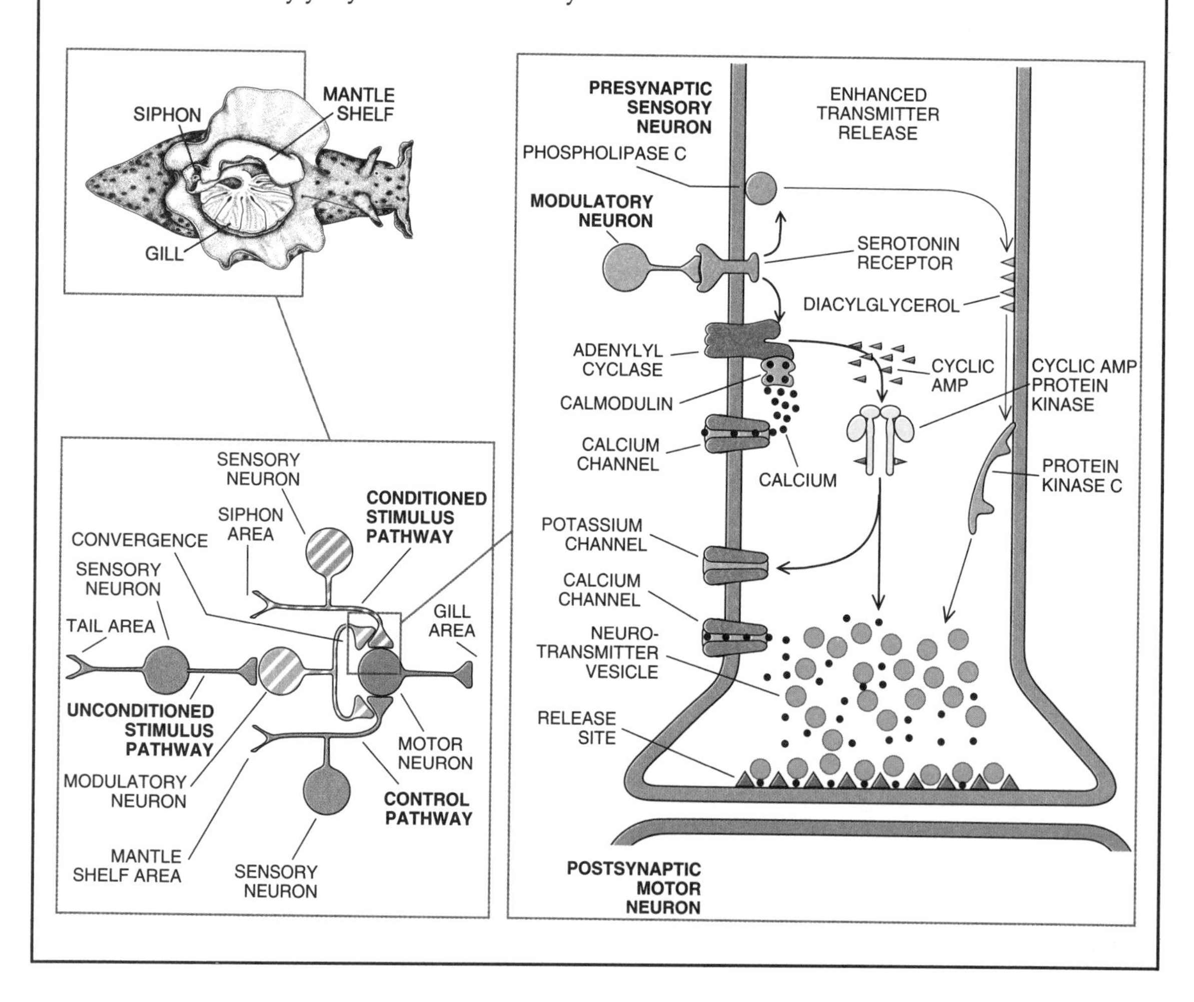

ulus enhance the action of serotonin? Action potentials produce a number of changes in the sensory neurons. They allow sodium and calcium to move in and potassium to move out, and they change the membrane potential. Abrams and Kandel found that the critical function of the action potential for activity dependence was the movement of calcium into the sensory neurons. Once in the cell, calcium binds to a protein called calmodulin, which amplifies the activation of the enzyme adenylyl cyclase by serotonin. When calcium/calmodulin binds to the adenylyl cyclase, the enzyme generates more cyclic AMP. This capacity makes adenylyl cyclase an important convergence site for the conditioned and the unconditioned stimuli.

Thus, the conditioned and the unconditioned stimuli are represented within the cell by the convergence of two different signals (calcium and serotonin) on the same enzyme. The 0.5-second interval between the two stimuli essential for learning in the gill-withdrawal reflex may correspond to the time during which calcium is elevated in the presynaptic terminal and binds to calmodulin so as to prime the adenylyl cyclase to produce more cyclic AMP in response to serotonin.

Activity-dependent amplification of the cyclic AMP pathway is not unique to the gill- or tail-withdrawal reflexes of *Aplysia*. Genetic studies in the fruit fly *Drosophila* have implicated a similar molecular mechanism for conditioning. *Drosophila* can be conditioned, and single-gene mutants have been discovered that are deficient in learning. One such mutant, called *rutabaga*, has been studied by William G. Quinn of the Massachusetts Institute of Technology and Margaret Livingstone of Harvard University and by Yadin Dudai of the Weizmann Institute in Israel. The gene encoding the defective protein in this mutant has now been shown to be a calcium/calmodulin-dependent adenylyl cyclase. As a result of the mutation in *rutabaga*, the cyclase has lost its ability to be stimulated by calcium/calmodulin. Moreover, Ronald L. Davis and his colleagues at Cold Spring Harbor Laboratory have found that this form of the adenylyl cyclase is enriched in the mushroom bodies, a part of the fly brain critical for several types of associative learning. Thus, both cell biological studies in *Aplysia* and genetic studies in *Drosophila* point to the significance of the cyclic AMP second-messenger system in certain elementary types of implicit learning and memory storage.

What about explicit forms of learning? Do these more complex types of associative learning also have cellular representations for associativity? If so, they must differ from the mechanisms for implicit learning because, unlike classical conditioning, explicit learning is often most successful when the two events that are associated occur simultaneously. For example, we recognize the face of an acquaintance most easily when we see that acquaintance in a specific context. The stimuli of the face and of the setting act simultaneously to help us recognize the person.

As we have seen, explicit learning in humans requires the temporal lobe. Yet it was unclear at first how extensive the bilateral lesion in the temporal lobe had to be to interfere with memory storage. Subsequent studies in humans and in experimental animals by Mortimer Mishkin of the National Institutes of Health and by Squire, David G. Amaral and Stuart Zola-Morgan of the University of California at San Diego help to answer the question. They suggest that one structure within the temporal lobe particularly critical for memory storage is the hippocampus. And yet lesions of the hippocampus interfere only with the storage of new memories: patients like H.M. still have a reasonably good memory of earlier events. The hippocampus appears to be only a temporary depository for long-term memory (see Figure 4.3). The hippocampus processes the newly learned information for a period of weeks to months and then transfers the information to relevant areas of the cerebral cortex for more permanent storage (see Chapter 5, "Brain and Language," by Antonio R. Damasio and Hanna Damasio). As discussed by Patricia S. Goldman-Rakic, the memory stored at these different cortical sites is then expressed through the working memory of the prefrontal cortex (see Chapter 6, "Working Memory and the Mind").

In 1973 Timothy Bliss and Terje Lømo, working in Per Andersen's laboratory in Oslo, Norway, first demonstrated that neurons in the hippocampus have remarkable plastic capabilities of the kind that would be required for learning. They found that a brief high-frequency train of action potentials in one of the neural pathways within the hippocampus produces an increase in synaptic strength in that pathway. The increase can be shown to last for hours in an anesthetized animal and for days and even weeks in an alert, freely moving animal.

Bliss and Lømo called this strengthening long-term potentiation (LTP). Later studies showed that LTP has different properties in different types of synapses within the hippocampus. We will focus here on an associative type of potentiation that has two interrelated characteristics. First, the associativity is of the Hebbian pre-post form: for facilitation to occur, the contributing presynaptic and postsynaptic neurons need to be active simultaneously. Second, and as a result, the long-term potentiation shows specificity: it is restricted in its action to the pathway that is stimulated.

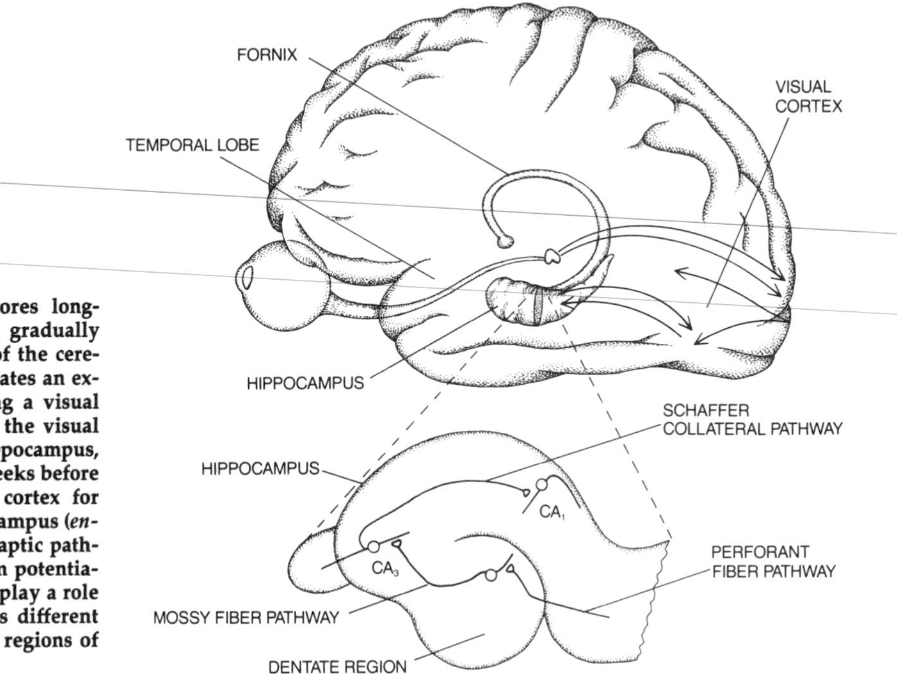

Figure 4.3 HIPPOCAMPUS stores long-term memory for weeks and gradually transfers it to specific regions of the cerebral cortex. The diagram illustrates an example of this process involving a visual image. Neural input travels to the visual cortex and then to the hippocampus, where it is stored for several weeks before it is transferred back to the cortex for long-term memory. The hippocampus (*enlargement*) has three major synaptic pathways, each capable of long-term potentiation (LTP), which is thought to play a role in the storage process. LTP has different properties in the CA_1 and CA_3 regions of the hippocampus.

Why is simultaneous firing of the presynaptic and postsynaptic cells necessary for long-term potentiation? The major neural pathways in the hippocampus use the amino acid glutamate as their transmitter. Glutamate produces LTP by binding to glutamate receptors on its target cells. It turns out that there are two relevant kinds of glutamate receptors: the NMDA receptors (named after the chemical *N*-methyl D-aspartate, which also binds to these receptors) and the non-NMDA receptors. Non-NMDA receptors dominate most synaptic transmission because the ion channel associated with the NMDA receptor is usually blocked by magnesium. It becomes unblocked only when the postsynaptic cell is depolarized. Moreover, optimal activation of the NMDA receptor channel requires that the two signals—glutamate binding to the receptor and depolarization of the postsynaptic cell—take place simultaneously. Thus, the NMDA receptor has associative or coincidence-detecting properties much as does the adenylyl cyclase. But its temporal characteristics, a requirement for simultaneous activation, are better suited for explicit rather than implicit forms of learning.

Calcium influx into the postsynaptic cell through the unblocked NMDA receptor channel is critical for long-term potentiation, as was first shown by Gary Lynch of the University of California at Irvine and by Roger A. Nicoll and Robert S. Zucker and colleagues at the University of California at San Francisco. Calcium initiates LTP by activating at least three different types of protein kinases.

The *induction* of LTP appears to depend on postsynaptic depolarization, leading to the influx of calcium and the subsequent activation of second-messenger kinases. For the *maintenance* of LTP, on the other hand, several groups of researchers have found that enhancement of transmitter from the presynaptic terminal is involved. These workers include Bliss and his colleagues, John Bekkers and Charles Stevens of the Salk Institute and Roberto Malinow and Richard Tsien of Stanford University.

If the induction of LTP requires a postsynaptic event (calcium influx through the NMDA receptor channels) and maintenance of LTP involves a presynaptic event (increase in transmitter release), then, as first proposed by Bliss, some message must be sent from the postsynaptic to the presynaptic neurons—and that poses a problem for neuroscientists. Ever since the great Spanish anatomist Santiago Ramón y Cajal first enunciated the principle of dynamic polarization, every chemical synapse studied has proved to be unidirectional. Information flows only from the presynaptic to the postsynaptic cell. In long-term potentiation, a new principle of nerve cell communication seems to be emerging. The calcium-activated second-messenger pathways, or perhaps calcium acting directly, seem to cause release of a retrograde plasticity factor from the active postsynaptic cell. This retrograde factor then diffuses to the presynaptic terminals to activate one or more second messengers that enhance transmitter release and thereby maintain LTP (see Figure 4.4).

Unlike the presynaptic terminals, which store transmitter in vesicles and release it at specialized release sites, the postsynaptic terminals lack any special release machinery. It therefore seemed attractive to posit that the retrograde messenger may be a substance that rapidly diffuses out of the post-

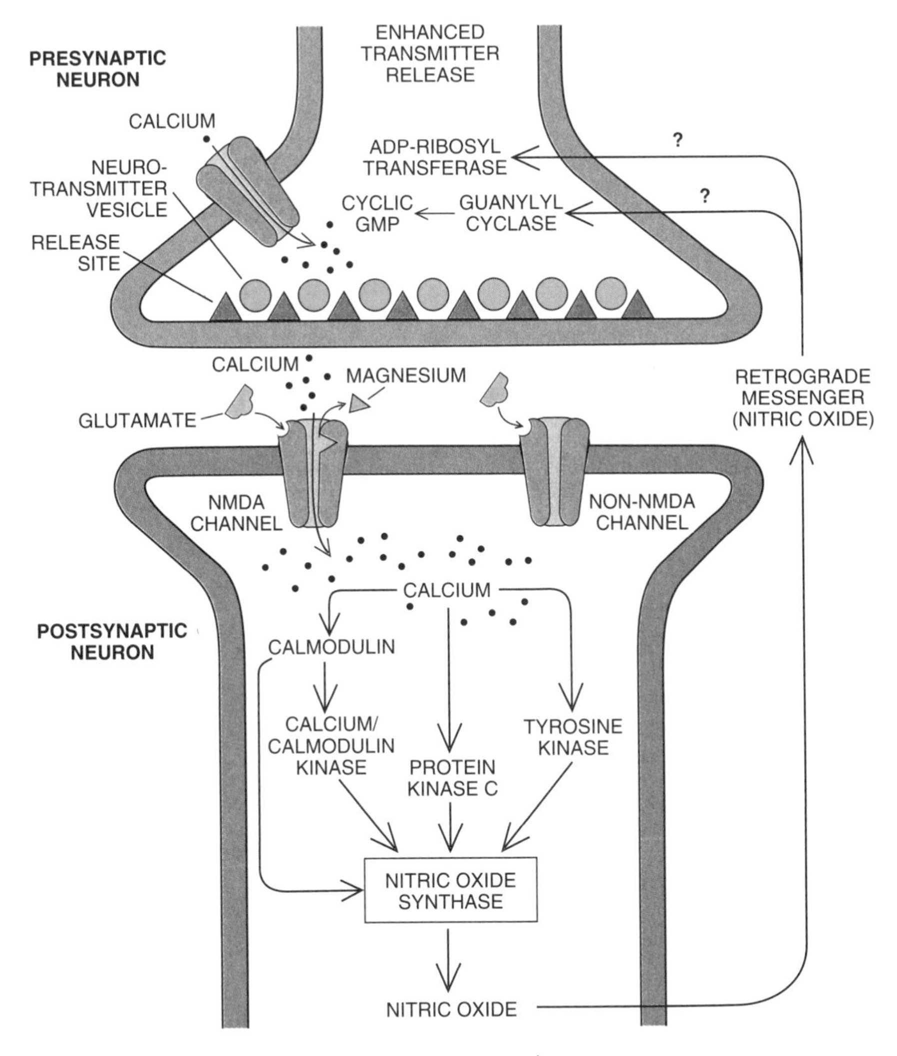

Figure 4.4 IN LONG-TERM POTENTIATION the postsynaptic membrane is depolarized by the actions of the non-NMDA receptor channels. The depolarization relieves the magnesium blockade of the NMDA channel, allowing calcium to flow through the channel. The calcium triggers calcium-dependent kinases that lead to the induction of LTP. The postsynaptic cell is thought to release a retrograde messenger capable of penetrating the membrane of the presynaptic cell. This messenger, which may be nitric oxide, is believed to act in the presynaptic terminal to enhance transmitter (glutamate) release, perhaps by activating guanylyl cyclase or ADP-ribosyl transferase.

synaptic cell across the synaptic cleft and into the presynaptic terminal. By 1991 four groups of researchers had obtained evidence that nitric oxide may be such a retrograde messenger: Thomas J. O'Dell and Ottavio Arancio in our laboratory, Erin M. Schuman and Daniel Madison of Stanford University, Paul F. Chapman and his colleagues at the University of Minnesota School of Medicine and Georg Böhme and his colleagues in France. Inhibiting the synthesis of nitric oxide in the postsynaptic neuron or absorbing nitric oxide in the extracellular space blocks the induction of LTP, whereas applying nitric oxide enhances transmitter release from presynaptic neurons.

In the course of studying the effects of applying nitric oxide in slices of hippocampus, we and Scott A. Small and Min Zhuo made a surprising finding: we discovered that nitric oxide produces LTP only if it is paired with activity in the presynaptic neurons, much as is the case in activity-dependent presynaptic facilitation in *Aplysia*. Presynaptic activity, and perhaps calcium influx, appears to be critical for nitric oxide to produce potentiation. These experiments suggest that long-term potentiation uses a combination of two independent, associative, synaptic learning mechanisms: a Hebbian NMDA receptor mechanism and a non-Hebbian, activity-dependent, presynaptic facilitating mechanism. According to this hypothesis, the activation of NMDA receptors in the postsynaptic cells produces a retrograde signal (nitric oxide). The signal then initiates an activity-dependent presynaptic mechanism, which facilitates the release of transmitter from the presynaptic terminals.

What might be the functional advantage of combining two associative cellular mechanisms, the postsynaptic NMDA receptor and the activity-dependent presynaptic facilitation, in this way? If presynaptic facilitation is produced by a diffusible substance, that substance could, in theory, find its way into neighboring pathways. In fact, studies by Tobias Bonhoeffer and his colleagues at the Max Planck Institute for Brain Research in Frankfurt indicate that LTP initiated in one postsynaptic cell spreads to neighboring postsynaptic cells. Activity dependence of presynaptic facilitation could be a way of ensuring that only specific presynaptic pathways—those that are active—are potentiated. Any inactive presynaptic terminals would not be affected (see Figure 4.5).

The changes in synapses that are thought to contribute to these instances of implicit and explicit learning raise a surprising reductionist possibility. The fact that associative synaptic changes do not require complex neural networks suggests there may be a direct correspondence between these associative forms of learning and basic cellular properties. In the cases that we have reviewed, the cellular properties seem to derive in turn from the properties of specific proteins—the adenylyl cyclase and the NMDA receptor—that are capable of responding to two independent signals, such as those from the conditioned stimulus and the unconditioned stimulus. Of course, these molecular associative mechanisms do not act in isolation. They are embedded in cells that have rich molecular machinery for elaborating the associative process. And the cells, in turn, are embedded in complex neural networks with considerable redundancy, parallelism and computational power, adding substantial complexity to these elementary mechanisms.

The finding that LTP occurs in the hippocampus, a region known to be significant in memory storage, made researchers wonder whether LTP is involved in the process of storing memories in this area of the brain. Evidence that it is has been provided by Richard Morris and his colleagues at the University of Edinburgh Medical School by means of a spatial memory task. When NMDA receptors in the hippocampus are blocked, the experimental animals fail to learn the task. These experiments suggest that NMDA receptor mechanisms in the hippocampus, and perhaps LTP, are involved in spatial learning.

Having now considered the mechanisms through which learning can produce changes in nerve cells, we are faced with a final set of questions. What are the mechanisms whereby the synaptic changes produced by explicit and implicit learning endure? How is memory maintained in the long term?

Experiments in both *Aplysia* and mammals indicate that explicit and implicit memory storage proceed in stages. Storage of the initial information, a type of short-term memory, lasts minutes to hours and involves changes in the strength of existing synaptic connections (by means of second-messenger-mediated modifications of the kind we have discussed). The long-term changes (those that persist for weeks and months) are stored at the same site, but they require something entirely new: the activation of genes, the expression of new proteins and the growth of new connections. In *Aplysia*, Craig H. Bailey, Mary C. Chen and Samuel M.

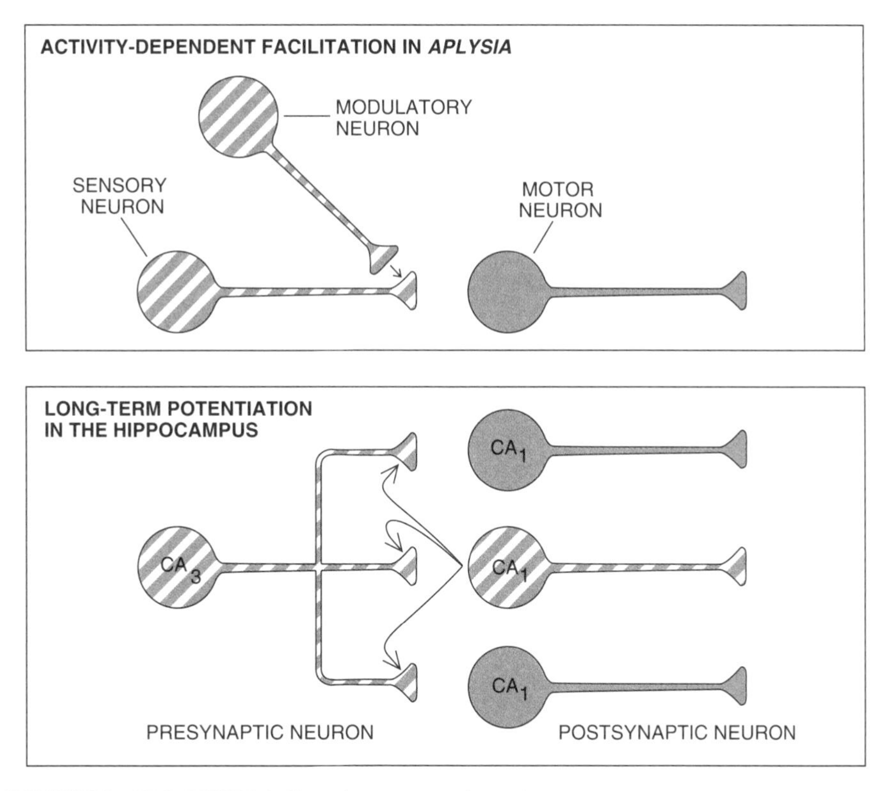

Figure 4.5 ASSOCIATIVE PROCESSES believed to contribute to learning in *Aplysia* and in the hippocampus of mammals may share similar mechanisms. Both may involve a modulatory substance that produces activity-dependent enhancement of transmitter release from the presynaptic neuron. Stripes denote neurons in which coincident activity must occur to produce the associative change.

Schacher and their colleagues at Columbia University and Byrne and his colleagues at the University of Texas Health Science Center have found that stimuli that produce long-term memory for sensitization and classical conditioning lead to an increase in the number of presynaptic terminals. Similar anatomic changes occur in the hippocampus after LTP.

If long-term memory leads to anatomic changes, does that imply that our brains are constantly changing anatomically as we learn and as we forget? Will we experience changes in our brain's anatomy as a result of reading and remembering this book?

This question has been addressed by many investigators, perhaps most dramatically by Michael Merzenich of the University of California at San Francisco. Merzenich examined the representation of the hand in the sensory area of the cerebral cortex. Until recently, neuroscientists believed this representation was stable throughout life. But Merzenich and his colleagues have now demonstrated that cortical maps are subject to constant modification based on use of the sensory pathways. Since all of us are brought up in somewhat different environments, are exposed to different combinations of stimuli and are likely to exercise our sensory and motor skills in different ways, the architecture of each of our brains will be modified in slightly different ways. This distinctive modification of brain architecture, along with a unique genetic makeup, contributes to the biological basis for the expression of individuality.

This view is best demonstrated in a study by Merzenich, in which he encouraged a monkey to touch a rotating disk with only the three middle fingers of its hand. After several thousand disk rotations, the area in the cortex devoted to the three middle fingers was expanded at the expense of that

Representation of the Surface of the Body in the Cortex

The homunculus ("little man") is a traditional way of illustrating how the surface of the body is represented in the somatosensory cortex. Larger areas of the cortex are devoted to parts of the body that have greater sensitivity, such as the fingers and lips. Recently the effects of sensitivity training have been shown in the owl monkey. The monkey's digits are represented in areas 3b and 1 of the somatosensory cortex (*a*). The diagrams (*b* and *d*) outline the regions that map the surface of the digits of an adult monkey (*c*) before and after training. During training the monkey rotated a disk for one hour a day, using only digits 2, 3 and occasionally 4. After three months of this activity, the area representing the stimulated fingers in the brain had increased substantially.

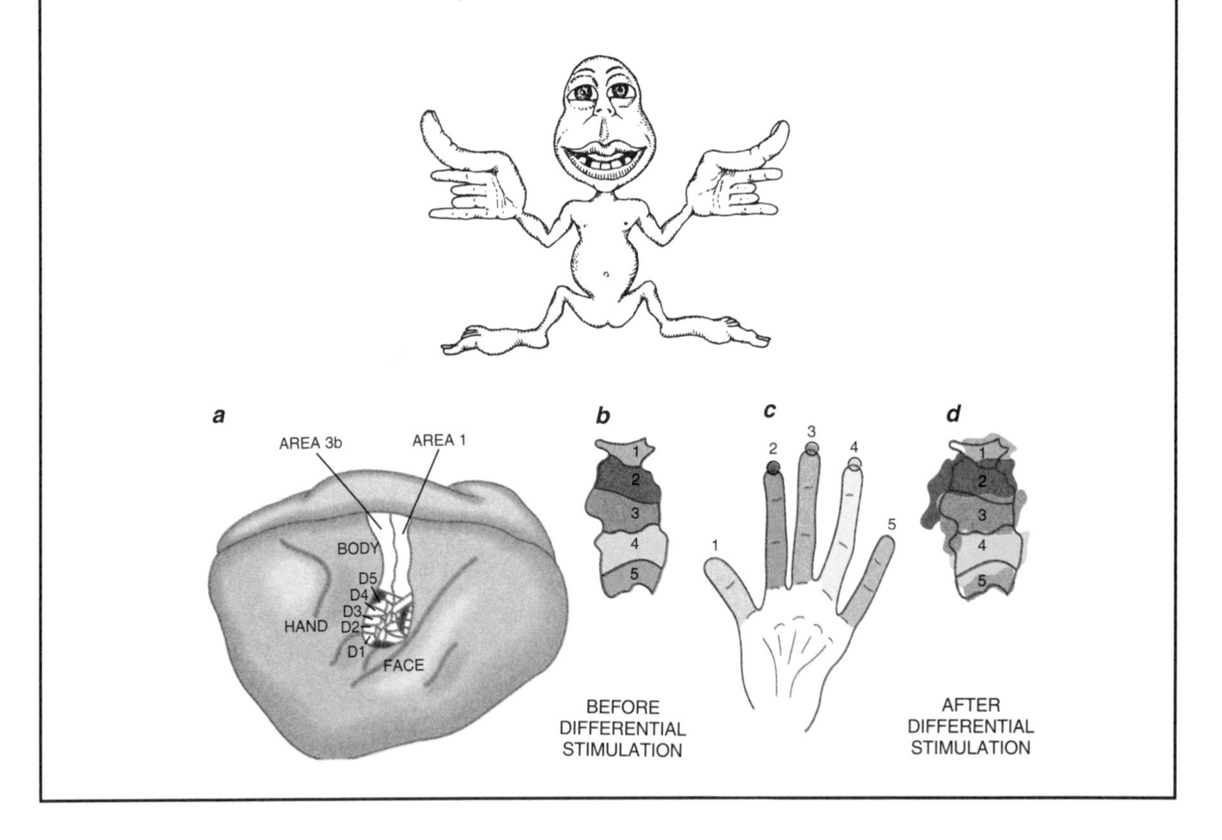

devoted to the other fingers (see boxed figure "Representation of the Surface of the Body in the Cortex"). Practice, therefore, can lead to changes in the cortical representation of the most active fingers. What mechanisms underlie the changes? Recent evidence indicates that the cortical connections in the somatosensory system are constantly being modified and updated on the basis of correlated activity, using a mechanism that appears similar to that which generates LTP.

Indeed, as we have learned from Carla J. Shatz (see Chapter 2, "The Developing Brain"), early results from cell biological studies of development suggest that the mechanisms of learning may carry with them an additional bonus. There is now reason to believe that the fine-tuning of connections during late stages of development may require an activity-dependent associative synaptic mechanism perhaps similar to LTP. If that is also true on the molecular level—if learning shares common molecular mechanisms with aspects of development and growth—the study of learning may help connect cognitive psychology to the molecular biology of the organism more generally. This broad biological unification would accelerate the demystification of mental processes and position their study squarely within the evolutionary framework of biology.

Brain and Language

A large set of neural structures serves to represent concepts; a smaller set forms words and sentences. Between the two lies a crucial layer of mediation.

• • •

Antonio R. Damasio and Hanna Damasio

What do neuroscientists talk about when they talk about language? We talk, it seems, about the ability to use words (or signs, if our language is one of the sign languages of the deaf) and to combine them in sentences so that concepts in our minds can be transmitted to other people. We also consider the converse: how we apprehend words spoken by others and turn them into concepts in our own minds (see Figure 5.1).

Language arose and persisted because it serves as a supremely efficient means of communication, especially for abstract concepts. Try to explain the rise and fall of the communist republics without using a single word. But language also performs what Patricia S. Churchland of the University of California at San Diego aptly calls "cognitive compression." It helps to categorize the world and to reduce the complexity of conceptual structures to a manageable scale.

The word "screwdriver," for example, stands for many representations of such an instrument, including visual descriptions of its operation and purpose, specific instances of its use, the feel of the tool or the hand movement that pertains to it. Or there is the immense variety of conceptual representations denoted by a word such as "democracy." The cognitive economies of language—its facility for pulling together many concepts under one symbol—make it possible for people to establish ever more complex concepts and use them to think at levels that would otherwise be impossible.

In the beginning, however, there were no words. Language seems to have appeared in evolution only after humans and species before them had become adept at generating and categorizing actions and at creating and categorizing mental representations of objects, events and relations. Similarly, infants' brains are busy representing and evoking concepts and generating myriad actions long before they utter their first well-selected word and even longer before they form sentences and truly use language. However, the maturation of language processes may not always depend on the maturation of conceptual processes, since some children with defective conceptual systems have nonetheless acquired grammar. The neural machinery necessary for some

Figure 5.1 MARTIN LUTHER KING is remembered both for his vision of racial harmony and for his ability to find words that stirred his listeners to action. The central issue of the neurophysiology of language, the authors say, is to map the structures in the brain that manipulate concepts and those that turn the concepts into words.

Components of a Sound-Based Language

PHONEMES	The individual sound units, whose concatenation, in a particular order, produces morphemes.
MORPHEMES	The smallest meaningful units of a word, whose combination creates a word. (In sign languages the equivalent of a morpheme is a visuomotor sign.)
SYNTAX	The admissible combinations of words in phrases and sentences (called grammar, in popular usage).
LEXICON	The collection of all words in a given language. Each lexical entry includes all information with morphological or syntactic ramifications but does not include conceptual knowledge.
SEMANTICS	The meanings that correspond to all lexical items and to all possible sentences.
PROSODY	The vocal intonation that can modify the literal meaning of words and sentences.
DISCOURSE	The linking of sentences such that they constitute a narrative.

syntactic operations seems capable of developing autonomously.

Language exists both as an artifact in the external world—a collection of symbols in admissible combinations—and as the embodiment in the brain of those symbols and the principles that determine their combinations. The brain uses the same machinery to represent language that it uses to represent any other entity. As neuroscientists come to understand the neural basis for the brain's representations of external objects, events and their relations, they will simultaneously gain insight into the brain's representation of language and into the mechanisms that connect the two.

We believe the brain processes language by means of three interacting sets of structures. First, a large collection of neural systems in both the right and left cerebral hemispheres represents nonlanguage interactions between the body and its environment, as mediated by varied sensory and motor systems—that is to say, anything that a person does, perceives, thinks or feels while acting in the world.

The brain not only categorizes these nonlanguage representations (along lines such as shape, color, sequence or emotional state), it also creates another level of representation for the results of its classification. In this way, people organize objects, events and relationships. Successive layers of categories and symbolic representations form the basis for abstraction and metaphor.

Second, a smaller number of neural systems, generally located in the left cerebral hemisphere, represent phonemes, phoneme combinations and syntactic rules for combining words (see boxed figure "Components of a Sound-Based Language"). When stimulated from within the brain, these systems assemble word-forms and generate sentences to be spoken or written. When stimulated externally by speech or text, they perform the initial processing of auditory or visual language signals.

A third set of structures, also located largely in the left hemisphere, mediates between the first two. It can take a concept and stimulate the production of word-forms, or it can receive words and cause the brain to evoke the corresponding concepts.

Such mediation structures have also been hypothesized from a purely psycholinguistic perspective. Willem J. M. Levelt of the Max Planck Institute for Psycholinguistics in Nijmegen has suggested that word-forms and sentences are generated from concepts by means of a component he calls "lemma," and Merrill F. Garret of the University of Arizona holds a similar view.

The concepts and words for colors serve as a particularly good example of this tripartite organization. Even those afflicted by congenital color blindness know that certain ranges of hue (chroma) band together and are different from other ranges, independent of their brightness and saturation. As Brent Berlin and Eleanor H. Rosch of the University of California at Berkeley have shown, these color concepts are fairly universal and develop whether or not a given culture actually has names to denote them. Naturally, the retina and the lateral geniculate nucleus perform the initial processing of color signals, but the primary visual cortex and at least two other cortical regions (known as V2 and V4) also participate in color processing; they fabricate what we know as the experience of color (see Figure 5.2).

With our colleague Matthew Rizzo, we have found that damage to the occipital and subcalcarine portions of the left and right lingual gyri, the region of the brain believed to contain the V2 and V4 cortices, causes a condition called achromatopsia. Patients who previously had normal vision lose their perception of color (see Chapter 3, "The Visual Image in Mind and Brain," by Semir Zeki). Furthermore, they lose the ability even to imagine colors. Achromatopsics usually see the world in shades of gray; when they conjure up a typically colored image in their minds, they see the shapes, movement and texture but not the color. When they think about a field of grass, no green is available, nor will red or yellow be part of their otherwise normal evocation of blood or banana. No lesion elsewhere in the brain can cause a similar defect. In some sense, then, the concept of colors depends on this region.

Patients with lesions in the left posterior temporal and inferior parietal cortex do not lose access to their concepts, but they have a sweeping impairment of their ability to produce proper word morphology regardless of the category to which a word belongs. Even if they are properly experiencing a given color and attempting to retrieve the corresponding word-form, they produce phonemically distorted color names; they may say "buh" for "blue," for example.

Other patients, who sustain damage in the temporal segment of the left lingual gyrus, suffer from a

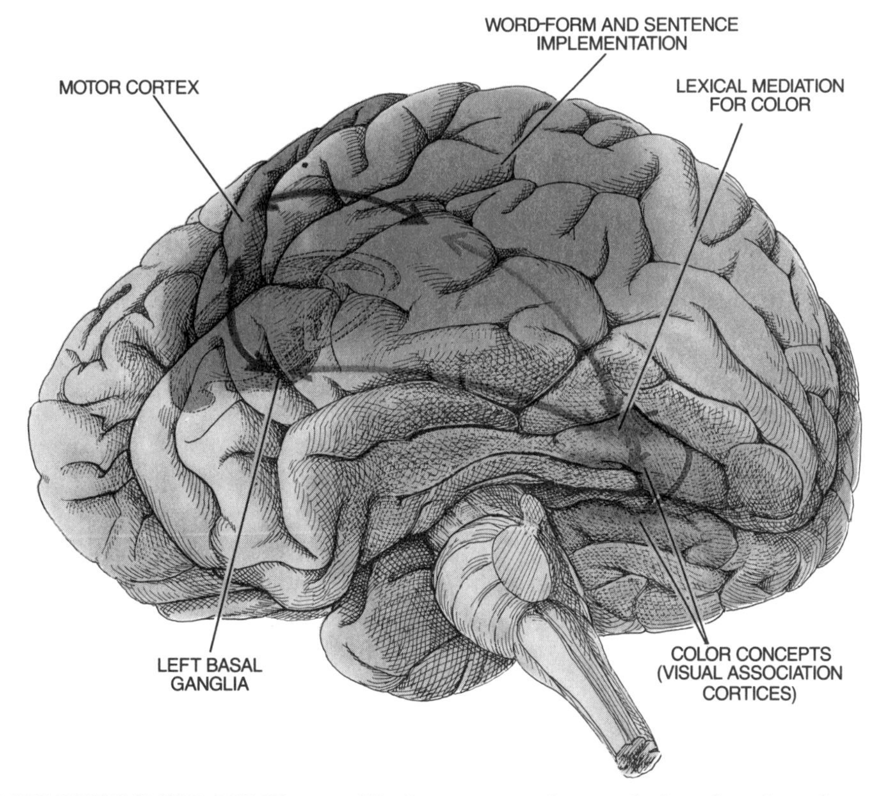

Figure 5.2 BRAIN SYSTEMS FOR COLOR exemplify the organization of language structures. Evidence from studies of people who have sustained brain damage indicates that the concepts for colors depend on the functioning of one system, that words for colors depend on another system and that the connections between words and concepts depend on a third.

peculiar defect called color anomia, which affects neither color concepts nor the utterance of color words. These patients continue to *experience* color normally: they can match different hues, correctly rank hues of different saturation and easily put the correct colored paint chip next to objects in a black-and-white photograph. But their ability to put names to color is dismally impaired. Given the limited set of color names available to those of us who are not interior decorators, it is surprising to see patients use the word "blue" or "red" when shown green or yellow and yet be capable of neatly placing a green chip next to a picture of grass or a yellow chip next to a picture of a banana. The defect goes both ways: given a color name, the patient will point to the wrong color.

At the same time, however, all the wrong color names the patient uses are beautifully formed, phonologically speaking, and the patient has no other language impairment. The color-concept system is intact, and so is the word-form implementation system. The problem seems to reside with the neural system that mediates between the two.

The same three-part organization that explains how people manage to talk about color applies to other concepts as well. But how are such concepts physically represented in the brain? We believe there are no permanently held "pictorial" representations of objects or persons as was traditionally thought. Instead the brain holds, in effect, a record of the neural activity that takes place in the sensory and motor cortices during interaction with a given object. The records are patterns of synaptic connections that can re-create the separate sets of activity that define an object or event; each record can also stimulate related ones. For example, as a person picks up a coffee cup, her visual cortices will respond to the colors of the cup and of its contents as well as to its shape and position. The somatosensory cortices will register the shape the hand assumes as it holds the cup, the movement of the hand and the arm as they bring the cup to the lips, the warmth of the coffee, and the body change some people call pleasure when they drink the stuff. The brain does not merely represent aspects of external reality; it also records how the body explores the world and reacts to it (see boxed figure "Components of a Concept").

The neural processes that describe the interaction between the individual and the object constitute a rapid sequence of microperceptions and microactions, almost simultaneous as far as consciousness is concerned. They occur in separate functional regions, and each region contains additional subdivisions: the visual aspect of perception, for example, is segregated within smaller systems specialized for color, shape and movement.

Where can the records that bind together all these fragmented activities be held? We believe they are embodied in ensembles of neurons within the brain's many "convergence" regions. At these sites of axons of feedforward projecting neurons from one part of the brain converge and join with reciprocally diverging feedback projections from other regions. When reactivation within the convergence zones stimulates the feedback projections, many anatomically separate and widely distributed neuron ensembles fire simultaneously and reconstruct previous patterns of mental activity.

In addition to storing information about experiences with objects, the brain also categorizes the information so that related events and concepts—shapes, colors, trajectories in space and time, and pertinent body movements and reactions—can be reactivated together. Such categorizations are denoted by yet another record in another convergence zone. The essential properties of the entities and processes in any interaction are thus represented in an interwoven fashion. The collected knowledge that can be represented includes the fact that a coffee cup has dimensions and a boundary; that it is made of something and has parts; that if it is divided it no longer is a cup, unlike water, which retains its identity no matter how it is divided; that it moved along a particular path, starting at one point in space and ending at another; that arrival at its destination produced a specific outcome. These aspects of neural representation bear a strong resemblance to the primitives of conceptual structure proposed by Ray Jackendoff of Brandeis University and the cognitive semantic schemas hypothesized by George P. Lakoff of the University of California at Berkeley, both working from purely linguistic grounds.

Activity in such a network, then, can serve both understanding and expression. The activity in the network can reconstruct knowledge so that a person experiences it consciously, or it can activate a system that mediates between concept and language, causing appropriately correlated word-forms and syntactical structures to be generated. Because the brain categorizes perceptions and actions simultaneously along many different dimensions, symbolic representations such as metaphor can easily emerge from this architecture.

Damage to parts of the brain that participate in these neural patterns should produce cognitive defects that clearly delineate the categories according to which concepts are stored and retrieved (the damage that results in achromatopsia is but one example of many). Elizabeth K. Warrington of the National Hospital for Nervous Diseases in London has studied category-related recognition defects and found patients who lose cognizance of certain classes of object. Similarly, in collaboration with our colleague Daniel Tranel, we have shown that access to concepts in a number of domains depends on particular neural systems.

For example, one of our patients, known as Boswell, no longer retrieves concepts for any unique entity (a specific person, place or event) with which he was previously familiar. He has also lost concepts for nonunique entities of particular classes. Many animals, for instance, are completely strange to him even though he retains the concept level that lets him know that they are living and animate. Faced with a picture of a raccoon, he says, "It is an animal," but he has no idea of its size, habitat or typical behavior.

Curiously, when it comes to other classes of nonunique entities, Boswell's cognition is apparently unimpaired. He can recognize and name objects, such as a wrench, that are manipulable and have a specific action attached to them. He can retrieve concepts for attributes of entities: he knows what it means for an object to be beautiful or ugly. He can grasp the idea of states or activities such as being in love, jumping or swimming. And he can understand abstract relations among entities or events such as "above," "under," "into," "from," "before," "after" or "during." In brief, Boswell has an impairment of concepts for many entities, all of which are denoted by nouns (common and proper). He has no problem whatsoever with concepts for attributes, states, activities and relations that are linguistically signified by adjectives, verbs, functors (prepositions, conjunctions and other verbal connective tissue) and syntactic structures. Indeed, the syntax of his sentences is impeccable.

Lesions such as Boswell's, in the anterior and middle regions of both temporal lobes, impair the brain's conceptual system. Injuries to the left hemisphere in the vicinity of the sylvian fissure, in contrast, interfere with the proper formation of words and sentences. This brain system is the most thoroughly investigated of those involved in language. More than a century and a half ago Paul Broca and Carl Wernicke determined the rough location of these basic language centers and discovered the phenomenon known as cerebral dominance—in most humans language structures lie in the left hemisphere rather than the right (see Figure 5.3). This disposition holds for roughly 99 percent of right-handed people and two thirds of left-handers. (The pace of research in this area has accelerated during the past two decades, thanks in large part to the influence of the late Norman Geschwind of Harvard Medical School and Harold Goodglass of the Boston Veterans Administration Medical Center.)

Studies of aphasic patients (those who have lost part or all of their ability to speak) from different language backgrounds highlight the constancy of these structures. Indeed, Edward Klima of the University of California at San Diego and Ursula Bellugi of the Salk Institute for Biological Studies in San Diego have discovered that damage to the brain's word-formation systems is implicated in sign-language aphasia as well. Deaf individuals who suffer focal brain damage in the left hemisphere can lose the ability to sign or to understand sign language. Because the damage in question is not to the visual cortex, the ability to see signs is not in question, just the ability to interpret them.

In contrast, deaf people whose lesions lie in the right hemisphere, far from the regions responsible for word and sentence formation, may lose conscious awareness of objects on the left side of their visual field, or they may be unable to perceive correctly spatial relations among objects, but they do not lose the ability to sign or understand sign language. Thus, regardless of the sensory channel through which linguistic information passes, the left hemisphere is the base for linguistic implementation and mediation systems.

Investigators have mapped the neural systems most directly involved in word and sentence formation by studying the location of lesions in aphasic patients. In addition, George A. Ojemann of the University of Washington and Ronald P. Lesser and Barry Gordon of Johns Hopkins University have stimulated the exposed cerebral cortex of patients undergoing surgery for epilepsy and made direct electrophysiological recordings of the response.

Damage in the posterior perisylvian sector, for

Components of a Concept

Concepts are stored in the brain in the form of "dormant" records. When these records are reactivated, they can re-create the varied sensations and actions associated with a particular entity or a category of entities. A coffee cup, for example, can evoke visual and tactile representations of its shape, color, texture and warmth, along with the smell and taste of the coffee or the path that the hand and the arm take to bring the cup from the table to the lips. All these representations are re-created in separate brain regions, but their reconstruction occurs fairly simultaneously.

example, disrupts the assembly of phonemes into words and the selection of entire word-forms. Patients with such damage may fail to speak certain words, or they may form words improperly ("loliphant" for "elephant"). They may, in addition, substitute a pronoun or a word at a more general taxonomic level for a missing one ("people" for "woman") or use a word semantically related to the concept they intend to express ("headman" for "president"). Victoria A. Fromkin of the University of California at Los Angeles has elucidated many of the linguistic mechanisms underlying such errors.

Damage to this region, however, does not disrupt patients' speech rhythms or the rate at which they speak. The syntactic structure of sentences is undisturbed even when there are errors in the use of functor words such as pronouns and conjunctions.

Damage to this region also impairs processing of speech sounds, and so patients have difficulty understanding spoken words and sentences. Auditory comprehension fails not because, as has been traditionally believed, the posterior perisylvian sector is a center to store "meanings" of words but rather because the normal acoustic analyses of the word-forms the patient hears are aborted at an early stage.

The systems in this sector hold auditory and kinesthetic records of phonemes and the phoneme sequences that make up words. Reciprocal projec-

tions of neurons between the areas holding these records mean that activity in one can generate corresponding activity in the other.

These regions connect to the motor and premotor cortices, both directly and by means of a subcortical path that includes the left basal ganglia and nuclei in the forward portion of the left thalamus. This dual motor route is especially important: the actual production of speech sounds can take place under the control of either a cortical or a subcortical circuit, or both. The subcortical circuit corresponds to "habit learning," whereas the cortical route implies higher-level, more conscious control and "associative learning" (see Chapter 4, "The Biological Basis of Learning and Individuality," by Eric R. Kandel and Robert D. Hawkins).

For instance, when a child learns the word-form "yellow," activations would pass through the word-formation and motor-control systems via both the cortical and subcortical routes, and activity in these areas would be correlated with the activity of the brain regions responsible for color concepts and mediation between concept and language. In time, we suspect, the concept-mediation system develops a direct route to the basal ganglia, and so the posterior perisylvian sector does not have to be strongly activated to produce the word "yellow." Subsequent

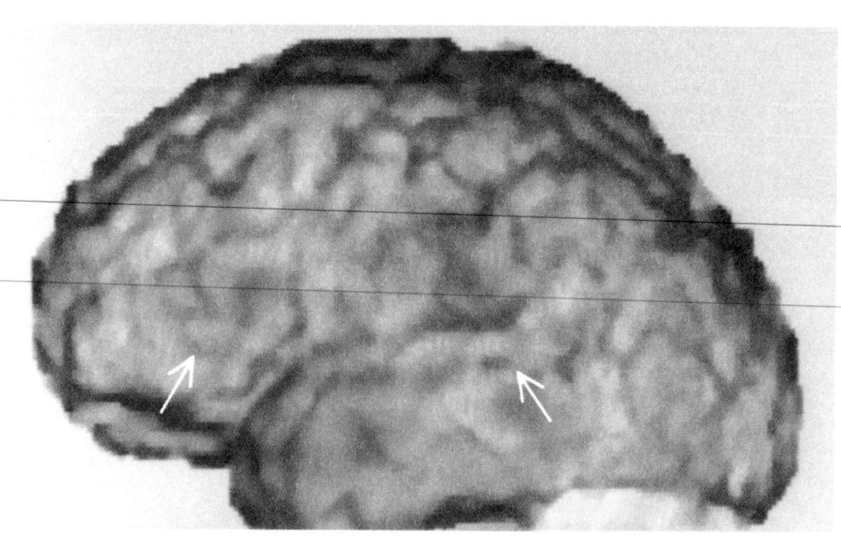

Figure 5.3 LANGUAGE ACTIVITY is visible in this positron emission tomographic (PET) scan of a normal individual performing a naming task. The PET image has been projected onto a three-dimensional magnetic resonance imaging (MRI) reconstruction of the same individual's brain. There are many areas of increased activity in the left hemisphere, including the motor cortex and the anterior and posterior language sectors (*arrows*). The image was produced by the University of Iowa's Department of Neurology, PET Facility and Image Analysis Facility.

learning of the word-form for yellow in another language would again require participation of the perisylvian region to establish auditory, kinesthetic and motor correspondences of phonemes.

It is likely that both cortical "associative" and subcortical "habit" systems operate in parallel during language processing. One system or the other predominates depending on the history of language acquisition and the nature of the item. Steven Pinker of the Massachusetts Institute of Technology has suggested, for example, that most people acquire the past tense of irregular verbs (take, took, taken) by associative learning and that of regular verbs (those whose past tense ends in *-ed*) by habit learning.

The anterior perisylvian sector, on the front side of the rolandic fissure, appears to contain structures that are responsible for speech rhythms and grammar. The left basal ganglia are part and parcel of this sector, as they are of the posterior perisylvian one. The entire sector appears to be strongly associated with the cerebellum; both the basal ganglia and the cerebellum receive projections from a wide variety of sensory regions in the cortex and return projections to motor-related areas. The role of the cerebellum in language and cognition, however, remains to be elucidated.

Patients with damage in the anterior perisylvian sector speak in flat tones, with long pauses between words, and have defective grammar. They tend in particular to leave out conjunctions and pronouns, and grammatical order is often compromised. Nouns come easier to patients with these lesions than do verbs, suggesting that other regions are responsible for their production.

Patients with damage in this sector have difficulty understanding meaning that is conveyed by syntactic structures. Edgar B. Zurif of Brandeis University, Eleanor M. Saffran of Temple University and Myrna F. Schwartz of Moss Rehabilitation Hospital in Philadelphia have shown that these patients do not always grasp reversible passive sentences such as "The boy was kissed by the girl," in which boy and girl are equally likely to be the recipient of the action. Nevertheless, they can still assign the correct meaning to a nonreversible passive sentence such as "The apple was eaten by the boy" or the active sentence "The boy kissed the girl."

The fact that damage to this sector impairs grammatical processing in both speech and understanding suggests that its neural systems supply the mechanics of component assembly at sentence level. The basal ganglia serve to assemble the components of complex motions into a smooth whole, and it seems reasonable that they might perform an analogous function in assembling word-forms into sentences. We also believe (based on experimental evidence of similar, although less extensive structures in monkeys) that these neural structures are closely interconnected with syntactic mediation units in the frontoparietal cortices of both hemispheres (see Figure 5.4). The delineation of those units is a topic of future research.

Between the brain's concept-processing systems and those that generate words and sentences lie the mediation systems we propose. Evidence for this neural brokerage is beginning to emerge from the study of neurological patients. Mediation systems not only select the correct words to express a particular concept, but they also direct the generation of sentence structures that express relations among concepts.

When a person speaks, these systems govern those responsible for word-formation and syntax; when a person understands speech, the word-formation systems drive the mediation systems. Thus far we have begun to map the systems that mediate proper nouns and common nouns that denote entities of a particular class (for example, visually ambiguous, nonmanipulable entities such as most animals).

Consider the patients whom we will call A.N. and L.R., who had sustained damage to the anterior and midtemporal cortices. Both can retrieve concepts normally: when shown pictures of entities or substances from virtually any conceptual category—human faces, body parts, animals and botanical specimens, vehicles and buildings, tools and utensils—A.N. and L.R. know unequivocally what they are looking at. They can define an entity's functions, habitats and value. If they are given

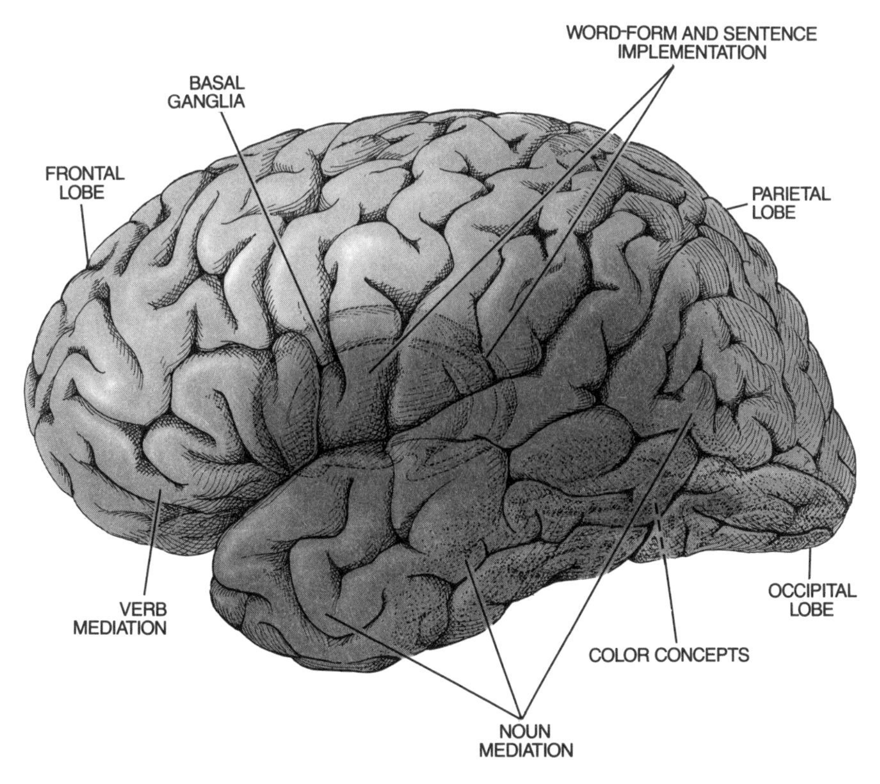

Figure 5.4 BRAIN SYSTEMS FOR LANGUAGE in the left hemisphere include word and sentence-implementation structures and mediation structures for various lexical items and grammar. The collections of neural structures that represent the concepts themselves are distributed across both right and left hemispheres in many sensory and motor regions.

sounds corresponding to those entities or substances (whenever a sound happens to be associated with them). A.N. and L.R. can recognize the item in question. They can perform this task even when they are blindfolded and asked to recognize an object placed in their hands.

But despite their obvious knowledge, they have difficulty in retrieving the names for many of the objects they know so well. Shown a picture of a raccoon, A.N. will say: "Oh! I know that it is—it is a nasty animal. It will come and rummage in your backyard and get into the garbage. The eyes and the rings in the tail give it away. I know it, but I cannot say the name." On the average they come up with less than half of the names they ought to retrieve. Their conceptual systems work well, but A.N. and L.R. cannot reliably access the word-forms that denote the objects they know.

The deficit in word-form retrieval depends on the conceptual category of the item that the patients are attempting to name. A.N. and L.R. make fewer errors for nouns that denote tools and utensils than for those naming animals, fruits and vegetables. (This phenomenon has been reported in similar form by Warrington and her colleague Rosaleen A. McCarthy of the National Hospital for Nervous Diseases and by Alfonso Caramazza and his colleagues at Johns Hopkins University.) The patients' ability to find names, however, does not split neatly at the boundary of natural and man-made entities. A.N. and L.R. can produce the words for such natural stimuli as body parts perfectly, whereas they cannot do the same for musical instruments, which are as artificial and as manipulable as garden tools.

In brief, A.N. and L.R. have a problem with the retrieval of common nouns denoting certain entities regardless of their membership in particular conceptual categories. There are many reasons why some entities might be more or less vulnerable to lesions than others. Of necessity, the brain uses different neural systems to represent entities that differ in structure or behavior or entities that a person relates to in different ways.

A.N. and L.R. also have trouble with proper nouns. With few exceptions, they cannot name friends, relatives, celebrities or places. Shown a picture of Marilyn Monroe, A.N. said, "Don't know her name but I know who she is; I saw her movies; she had an affair with the president; she committed suicide; or maybe somebody killed her; the police, maybe?" These patients do not have what is known as face agnosia or prosopagnosia—they can recognize a face without hesitation—but they simply cannot retrieve the word-form that goes with the person they recognize.

Curiously, these patients have no difficulty producing verbs. In experiments we conducted in collaboration with Tranel, these patients perform just as well as matched control subjects on tasks requiring them to generate a verb in response to more than 200 stimuli depicting diverse states and actions. They are also adept at the production of prepositions, conjunctions and pronouns, and their sentences are well formed and grammatical. As they speak or write, they produce a narrative in which, instead of the missing noun, they will substitute words like "thing" or "stuff" or pronouns such as "it" or "she" or "they." But the verbs that animate the arguments of those sentences are properly selected and produced and properly marked with respect to tense and person. Their pronunciation and prosody (the intonation of the individual words and the entire sentence) are similarly unexceptionable.

The evidence that lexical medication systems are confined to specific regions is convincing. Indeed, the neural structures that mediate between concepts and word-forms appear to be graded from back to front along the occipitotemporal axis of the brain. Mediation for many general concepts seems to occur at the rear, in the more posterior left temporal regions; mediation for the most specific concepts takes place at the front, near the left temporal pole. We have now seen many patients who have lost their proper nouns but retain all or most of their common nouns. Their lesions are restricted to the left temporal pole and medial temporal surface of the brain, sparing the lateral and inferior temporal lobes. The last two, in contrast, are always damaged in the patients with common noun retrieval defects.

Patients such as A.N. and L.R., whose damage extends to the anterior and midtemporal cortices, miss many common nouns but still name colors quickly and correctly. These correlations between lesions and linguistic defects indicate that the temporal segment of the left lingual gyrus supports mediation between color concepts and color names, whereas mediation between concepts for unique persons and their corresponding names requires neural structures at the opposite end of the network, in the left anterior temporal lobe. Finally, one of our more recent patients, G.J., has extensive damage that encompasses all of these left occipito-temporal regions from front to back. He has lost access to a sweeping universe of noun word-forms and is equally unable to name colors or unique persons. And yet his concepts are preserved. The results in these patients support Ojemann's finding of impaired language processing after electrical stimulation of cortices outside the classic language areas.

It appears that we have begun to understand fairly well where nouns are mediated, but where are the verbs? Clearly, if patients such as A.N. and L.R. can retrieve verbs and functor words normally, the regions required for those parts of speech cannot be in the left temporal region. Preliminary evidence points to frontal and parietal sites. Aphasia studies performed by our group and by Caramazza and Gabriele Miceli of Catholic University of the Sacred Heart, Milan, and Rita Berndt of the University of Maryland show that patients with left frontal damage have far more trouble with verb retrieval than with noun retrieval.

Additional indirect evidence comes from positron emission tomography (PET) studies conducted by Steven E. Petersen, Michael I. Posner and Marcus E. Raichle of Washington University. They asked research subjects to generate a verb corresponding to the picture of an object—for example, a picture of an apple might generate "eat." These subjects activated a region of the lateral and inferior dorsal frontal cortex that corresponds roughly to the areas delineated in our studies. Damage to these regions not only compromises access to verbs and functors, it also disturbs the grammatical structure of the sentences that patients produce.

Although this phenomenon may seem surprising at first, verbs and functors constitute the core of syntactic structure, and so it makes sense that the mediation systems for syntax would overlap with them. Further investigations, either of aphasic patients or of normal subjects, whose brain activity can be mapped by PET scans, may clarify the precise arrangement of these systems and yield maps like those that we have produced to show the differing locations of common and proper nouns.

During the past two decades, progress in understanding the brain structures responsible for language has accelerated significantly. Tools such as magnetic resonance imaging have made it possible to locate brain lesions accurately in patients suffering from aphasia and to correlate specific language deficits with damage to particular regions of the brain. And PET scans offer the opportunity to study the brain activity of normal subjects engaged in linguistic tasks.

Considering the profound complexity of language phenomena, some may wonder whether the neural machinery that allows it all to happen will ever be understood. Many questions remain to be answered about how the brain stores concepts. Mediation systems for parts of speech other than nouns, verbs and functors, have been only partially explored. Even the structures that form words and sentences, which have been under study since the middle of the 19th century, are only sketchily understood.

Nevertheless, given the recent strides that have been made, we believe these structures will eventually be mapped and understood. The question is not if but when.

Working Memory and the Mind

Anatomic and physiological studies of monkeys are locating the neural machinery involved in forming and updating internal representations of the outside world. Such representations form a cornerstone of the rational mind.

• • •

Patricia S. Goldman-Rakic

The seeming simplicity of everyday life belies the enormously complex ongoing operations of the mind. Even routine tasks such as carrying on a conversation or driving to work draw on a mixture of current sensory data and stored knowledge that has suddenly become relevant. The combination of moment-to-moment awareness and instant retrieval of archived information constitutes what is called the working memory, perhaps the most significant achievement of human mental evolution. It enables humans to plan for the future and to string together thoughts and ideas, which has prompted Marcel Just and Patricia Carpenter of Carnegie Mellon University to refer to working memory as "the blackboard of the mind."

Until recently, the fundamental processes involved in such higher mental functions defied description in the mechanistic terms of science. Indeed, for the greater part of this century, neurobiologists often denied that such functions were accessible to scientific analysis or declared that they belonged strictly to the domain of psychology and philosophy. Within the past two decades, however, neuroscientists have made great advances in understanding the relation between cognitive processes and the anatomic organization of the brain. As a consequence, even global mental attributes such as thought and intentionality can now be meaningfully studied in the laboratory.

The ultimate goal of that work is extraordinarily ambitious. Eventually researchers such as myself hope to be able to analyze higher mental functions in terms of the coordinated activation of neurons in various structures in the brain. It should also be possible to identify the cells that mediate the activity of those structures. Such research will help explain the origin of mind. It may also lead to more complete descriptions of baffling mental disorders such as schizophrenia.

For many years, insight into the operation of the brain was stymied by the misconception that memory is a single entity that could be traced to a single structure or location. Since the 1950s, neuroscientists have increasingly come to appreciate that memory consists of multiple components constructed around a distributed network of neurons.

Figure 6.1 WORKING MEMORY enables a human to retrieve stored symbolic information, such as the bowings and fingerings of a memorized piece of music, and to translate that information into a controlled set of motor activities. Studies of similar but simpler information processing performed by primates is revealing the structure of working memory.

According to present thinking, a form of memory known as associative memory acquires facts and figures and holds them in long-term storage. That knowledge is of no use, however, unless it can be accessed and brought to mind in order to influence current behavior.

Working memory complements associative memory by providing for the short-term activation and storage of symbolic information, as well as by permitting the manipulation of that information (see Figure 6.1). A simple activity involving working memory is the carry-over operation in mental arithmetic, which requires temporarily storing a string of numbers and holding the sum of one addition in mind while calculating the next. More complex examples include planning a chess move or constructing a sentence. Working memory in humans is considered fundamental to language comprehension, to learning and to reason.

Numerous lines of evidence indicate that the operations of working memory are carried out in a part of the brain known as the prefrontal lobes of the cerebral cortex. (Cortex derives from the Latin word meaning bark; the cerebral cortex consists of an outer rind of so-called gray matter neurons surrounding the cerebrum.) Much of the evidence identifying this structure as the center for working memory comes from observations of the effects of injuries to the prefrontal part of the hemispheres. For example, patients having frontal lobe damage exhibit gross deficiencies in how they use knowledge to guide their behavior in everyday situations. Nevertheless, they often retain a full store of information and may continue to score well on conventional tests of intelligence.

Although most fully developed in humans, some elements of working memory exist in other animals, especially in other primates; if their prefrontal cortices are damaged, those animals develop symptoms much like the ones seen in humans. Neuroscientists have therefore turned to monkeys in their efforts to explore the nature of working memory. Such exploration has been aided by the design of repeatable tests of working memory functions.

Working memory is being assessed in monkeys by means of tasks known as delayed-response tests, which evaluate an organism's ability to react to situations on the basis of stored or internalized representations rather than on information immediately present in the environment. In the prototypical delayed-response test, an animal receives a brief visual or auditory stimulus that is then hidden or taken away (see Figure 6.2). After a delay of several seconds, the animal is given a signal that tells it to respond to the location where the stimulus had appeared. If the response is correct, the animal receives a reward, usually food or juice.

Delayed-response tests tap working memory processes because the animal must retain the memory of the location of the stimulus during the period of the delay. The proper response at the end of the delay is indicated not by external stimuli but by the memory of what the subject saw on the previous trial. Furthermore, the correct response may differ from one trial to the next, depending on new information presented to the subject in each trial. Correct responses in working memory tasks, as in human affairs, are guided by memory rather than by immediate sensory information, and they depend on constant updating of the relevant information.

Delayed-response tests resemble very closely the object-permanence task, developed in the early part of this century by the French child psychologist Jean Piaget, that is widely used to chart the cognitive development of young children. For Piaget's task, a child is shown two boxes, one of which contains a toy. The boxes are then closed. After a brief wait, during which the child is purposely distracted, the child is asked to pick which box contains the toy. Once the child gives several consecutive correct responses, the toy is switched into the other box while the child watches. The experimenter then continues the test to find out whether the child will change his or her response in accord with the updated information.

A series of studies has demonstrated that performance on the object-permanence task, like the ability to conduct delayed-response activities, depends on the degree of maturity of the subject's prefrontal cortex. Human infants less than about eight months old (whose cortices have not yet acquired adult circuitry) perform poorly on these tasks, as do monkeys whose prefrontal regions have been surgically ablated. In both cases, the subjects' responses are guided by habit and by reflex rather than by representational principles. Infants and brain-injured monkeys tend to repeat the response that previously was reinforced (for example, choosing the box on the right even after they have seen that the toy was transferred to the box on the left) rather than change their response to agree with newly presented information. Both humans and

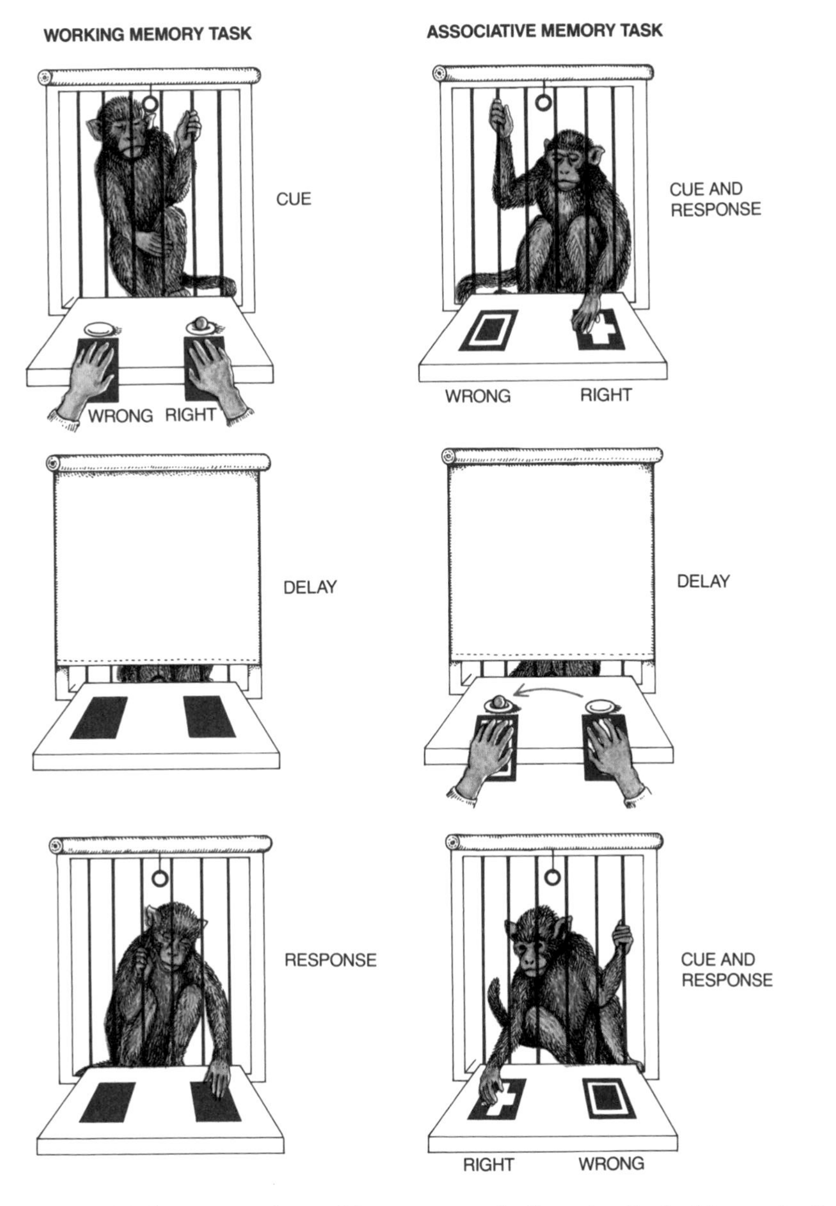

Figure 6.2 MEMORY TASKS help to assess the workings of the mind. In the classical working memory task (*left*), a monkey briefly views a target stimulus—in this case, a morsel of food. Only after a delay is the animal allowed to retrieve the food. The experimenter randomly varies the location of the food between trials, so that each response tests only the animal's short-term retention of visual and spatial information. An associative memory task (*right*), in contrast, follows a consistent pattern throughout. Here a plus sign always indicates the correct response. The task therefore measures the animal's ability to retain long-term rules.

monkeys act as if "out of sight" is "out of mind."

Such behavior implies that the mechanism for guiding behavior by representational knowledge is destroyed in monkeys having prefrontal lesions and not yet developed in human infants. In support of that notion, I, along with Jean-Pierre Bourgeois and Pasko Rakic, also at Yale University, have examined the rate at which neural connections form in the prefrontal cortices of juvenile monkeys.

The time of most rapid synapse formation in the

animals' prefrontal region occurs when the animals are roughly two to four months old, the same age at which the monkeys acquire the capacity to perform delayed-response tasks. The concept that an object exists continuously in space and time even when out of view and, more generally, the ability to form abstract concepts may depend on a fundamental capacity to store representations of the outside world and to respond to those representations even when the real objects are not present.

The studies described above raised the enticing possibility of identifying more precisely the brain structures associated with delayed-response activities and representational memory. Much of the progress toward that goal has derived from experiments that monitor electrical activity in single neurons in monkeys' prefrontal cortices while the animals perform tasks that depend on specific delayed-response skills.

Joaquin M. Fuster of the University of California at Los Angeles, along with Kisou Kubota and Hiroaki Niki of the Kyoto Primate Center in Japan, performed the first experiments of how individual neurons behave in the prefrontal cortex. The researchers introduced fine electrodes into the prefrontal cortices of monkeys trained to perform simple delayed-response tasks and then recorded the animals' neuronal activity in relation to the events in the task. Those studies revealed a range of responses among the neurons in the prefrontal cortex. Some cells showed heightened electrical activity when information was presented, whereas others became active during the delay period, when the animals were remembering the information. A third set of neurons responded most strongly when the animals began their motor response.

At Yale, Shintaro Funahashi, Charles J. Bruce and I have used the single-neuron technique in conjunction with a delayed-response experiment that tests spatial memory (see Figure 6.3). For our experiment, a monkey is trained to fix its gaze on a small spot in the center of a television screen. A visual stimulus, typically a small square, appears briefly in one of eight locations on the screen and then vanishes. At the end of a delay of three to six seconds, the central light, or fixation spot, switches off, instructing the animal to move its eyes to the location where the stimulus was seen before the delay. If the response is correct, the animal is rewarded with a sip of grape juice. Because the animal's gaze is locked onto the fixation spot, each stimulus activates a specific set of retinal cells. Those cells, in turn, trigger only a certain subset of the visual pathways in the brain.

Using the eye-movement experiment, we have demonstrated that certain neurons in the prefrontal cortex possess what we call "memory fields": when a particular target disappears from view, an individual prefrontal neuron switches into an active state, producing electrical signals at more than twice the baseline rate. The neuron remains activated until the end of the delay period, when the animal delivers its response. A given neuron appears always to code the same visual location. For example, some neurons fire only if the stimulus appears at the nine o'clock position on the television screen; the cell does not respond to visual stimuli that appear elsewhere in the monkey's visual field. Other neurons code for other target locations in working memory.

The neurons capable of retaining the visual and spatial coordinates of a stimulus (in other words, of keeping its location "in mind" after it vanishes) appear to be organized together within a specific area of the prefrontal cortex. These neurons collectively form the core of the spatial working memory system. If the activity of one or more of these neurons falters during the delay period—if the animal is distracted, for example—the animal will probably make an error.

The activation of prefronal neurons during the delay period of a delayed-response task depends neither on the presence of an external stimulus nor on the execution of a response. Rather the neural activity corresponds to a mental event interposed between the stimulus and the response. Monkeys whose prefrontal cortices have been damaged have no difficulty in moving their eyes to a visible target or in reaching for a desired object, but they cannot direct those motor responses by remembering targets and objects that are no longer in evidence.

Because the prefrontal cortex functions as an intermediary between memory and action, one can imagine that damage to the prefrontal cortex could spare knowledge about the outside world yet destroy the organism's ability to bring that stored knowledge to mind and to utilize it. Indeed, monkeys whose prefrontal cortices have been damaged, as well as many humans with similar injuries, exhibit no difficulty learning sensory-discrimination tasks. All forms of associative, or long-term, learn-

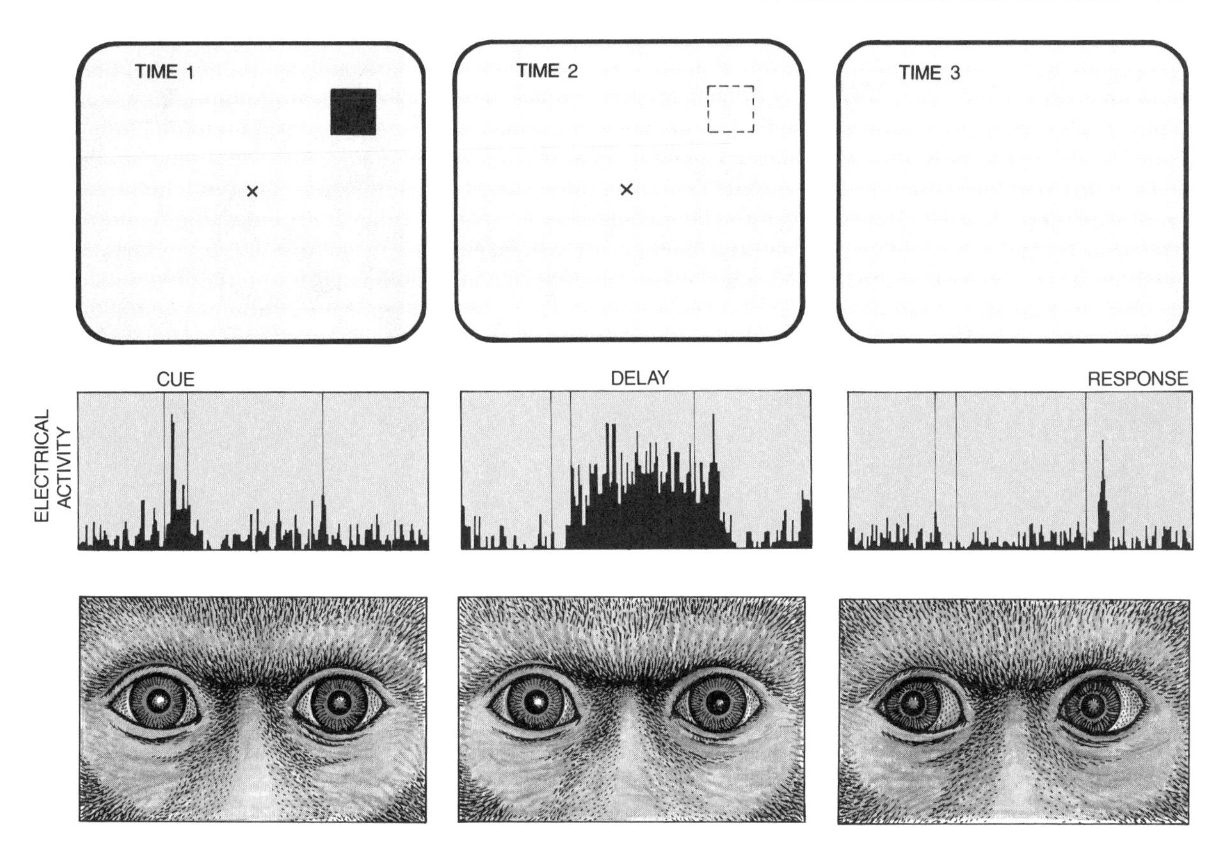

Figure 6.3 DELAYED-RESPONSE TASK has been used to study working memory in monkeys. While a monkey fixes its gaze on a central spot, a target flashes on the screen (*left*), then vanishes. During a delay of several seconds, the monkey keeps a memory of the spot "in mind" (*center*). When the central spot turns off, the animal moves its eyes to look where the target appeared (*right*). Measurements of electrical activity show that certain neurons in the prefrontal cortex react to the appearance of the target, others hold the memory of it in mind and still others fire in preparation for a motor response.

ing are preserved as long as the subject can still find the familiar environmental stimuli associated with certain consequences and expectations (see Chapter 4, "The Biological Basis of Learning and Individuality," by Eric R. Kandel and Robert D. Hawkins).

Over the past decade, improved techniques for investigating the anatomy of the brain have provided for the first time an accurate and detailed picture of how the prefrontal cortex connects with major sensory and motor control centers. Various researchers have found that the part of the cortex near the principal sulcus, a large groove in the prefrontal cortex, is critical for the visual and spatial working memory functions. I have focused my research on this particular region in the belief that an in-depth neurobiological analysis of one major subdivision of the prefrontal cortex could serve as a starting point for analysis of the other subdivisions of the brain and help lead the way to development of a unified theory of the function of the entire prefrontal cortex.

Studies of direct and indirect neuronal linkages in the brain reveal that the prefrontal cortex is part of an elaborate network of reciprocal connections between the principal sulcus and the major sensory, limbic and premotor areas of the cerebral cortex (see Figure 6.4). That particular network seems to be dedicated to spatial information processing. The network's structure probably follows the same basic plan as do other similarly organized networks that draw on multiple parts of the brain and are dedicated to other cognitive functions—object recogni-

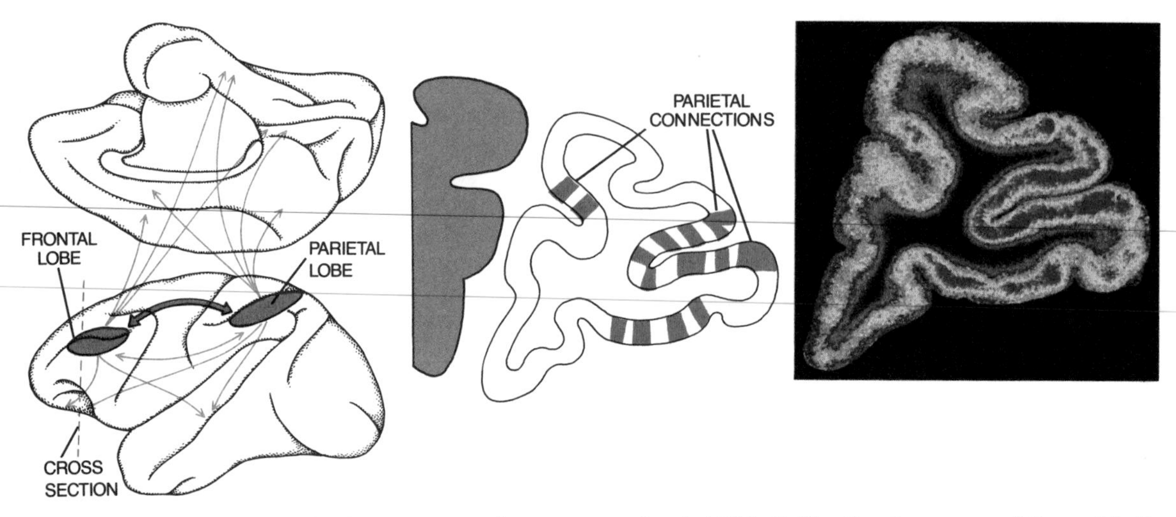

Figure 6.4 NEURONAL CIRCUITRY connects the prefrontal cortex to the sensory, limbic and motor systems in a monkey brain (*top*). Anatomic studies show that neural projections from the parietal lobe to the prefrontal cortex exhibit a modular pattern, as seen in this frontal cross section (*middle*). Radioactive tracers reveal the metabolic activity in a frontal cross section of the brain of a monkey performing a delayed-response task (*bottom*). The distribution of activity closely conforms to the anatomic links.

tion, language production and comprehension, and mathematical reasoning, for example.

As previously noted, delay-response experiments demonstrate that neurons in the principal sulcus are sensitive to the specific location of visual stimuli. Those neurons must therefore have access to visual and spatial information originating elsewhere in the brain. The principal sulcus does in fact receive signals from the posterior parietal cortex, where the brain processes spatial vision. Clinical studies have documented that damage to the parietal cortex in humans causes spatial neglect, a loss of awareness of the body and its relation to objects in the outside world.

Given that working memory depends on accessing and bringing to mind information that is stored in long-term memory, one might presume that the principal sulcus also interacts with the hippocampus, the neuronal structure that controls associative, or learned, memory. Researchers have used radioactive amino acids to trace direct connections between the principal sulcus and the hippocampus.

My colleague Harriet Friedman, also at Yale, and I have used a remarkable technique known an autoradiography to measure brain metabolism. Our work shows that the hippocampus and the principal sulcal areas of the cortex are often simultaneously active during delayed-response tests. My coworkers and I think that the primary role of the hippocampus is to consolidate new associations, whereas the prefrontal cortex is necessary for retrieving the products of such associative learning (facts, events, rules) from long-term storage elsewhere in the brain for use in the task at hand.

A particularly useful version of autoradiography, called the 2-deoxyglucose method, has made it possible to observe directly which parts of the brain are activated during specific tasks. In this technique, developed by Louis Sokoloff of the National Institute of Mental Health, animals are injected with the compound 2-deoxyglucose, a molecule that appears chemically identical to glucose, the sugar that cells consume to provide energy. The more active a cell is, the more 2-deoxyglucose it takes in. Unlike normal glucose, however, 2-deoxyglucose cannot be broken down by metabolic activity, so it accumulates in the cell. Sokoloff uses a radioactive version of the compound. The concentration of radioactivity in each part of the brain is therefore directly proportional to how active the cells there have been.

For our studies, a monkey trained to perform the delayed-response tasks receives an intravenous injection of radioactive 2-deoxyglucose. Immediately after completing the task, the animal is sacrificed and its brain is dissected into thin slices that are placed on photographic film. Radioactivity darkens the film, so each exposure serves as a snapshot of the activity of the cells in one particular slice of the brain.

My colleagues and I have found that the prefrontal cortex, as well as many of the areas with which it is connected (for example, the hippocampus, the bottom portion of the parietal cortex and the thalamus), exhibits a high level of metabolic activity during delayed-response performance. The same areas are notably less active when the monkey performs associative memory tasks that do not depend on short-term, rapid updating of information.

These results confirm anatomic studies of the connections between the prefrontal cortex and other parts of the brain. More significantly, they also reveal the degree to which various parts of the brain are engaged in certain discrete memory tasks. The studies also hint at how the prefrontal cortex organizes the many different kinds of information that must flood through it. In fact, patterns of brain activity appear distinctly different depending on whether the task calls up memories of location or of attributes of objects.

I think the prefrontal cortex is divided into multiple memory domains, each specialized for encoding a different kind of information, such as the location of objects, the features of objects (color, size and shape) and additionally, in humans, semantic and mathematical knowledge. Recently Fraser Wilson and James Skelly in my laboratory at Yale have begun to define an area below the principal sulcus in monkeys where neurons respond preferentially to complex attributes of objects rather than to their locations. They have found neurons there that increase their rate of firing when a monkey is remembering a red circle but not when calling up a memory of a green square, for example.

Noninvasive imaging techniques are increasingly being used to monitor activation patterns in the human brain and to identify which neurons are engaged during specific mental tasks. One, known as positron emission tomography (PET), resembles autoradiography in that the subject takes in a radioactive compound that exposes changes in blood flow to a given region of the brain, indirectly displaying that region's degree of metabolic activity. Another way to record human brain activity is to measure the changing electrical potentials on the scalp in response to controlled sensory stimulation, a procedure called electroencephalography (EEG). Neither PET scans nor EEGs can provide anything close to the resolution possible in 2-deoxyglucose studies in animals, but they are invaluable tools for monitoring the human brain during mental activity.

A series of PET studies at Hammersmith Hospital in London and at Washington University examined subjects performing tasks that required them to keep a mental record of recently presented lists of words. Another PET experiment by the Washington University group required subjects to generate an appropriate verb to accompany a noun flashed in front of them on a card. The participants in all three tests displayed heightened neuronal activity in the prefrontal cortex while performing their tasks, all of which engaged working memory.

In a complementary study, Robert T. Knight of the University of California at Davis looked at EEGs of patients whose frontal lobes were injured. He asked them to perform tasks that depended on comparing current auditory stimuli with recently presented ones in order to detect whether they are the same or different. Frontal lobe patients displayed patterns of electrical activity quite unlike those of healthy subjects performing the same task, suggesting that the patients do not store recent information in memory the same way as do normal people.

In one study, subjects were exposed to steady patterns of low and high tones and occasional, unexpected auditory stimuli. Healthy people developed positive electrical potentials on their cortices within one third of a second of hearing the anomalous sound. Patients who had lesions in their prefrontal cortices showed no such response, although they reacted normally to the familiar background tones. These data are consistent with the notion that the prefrontal cortex temporarily stores information against which current stimuli are judged.

The ultimate function of the neurons in the prefrontal cortex is to excite or inhibit activity in other parts of the brain. In this way, information processed in the principal sulcus can direct neurons in the motor centers that in turn carry out movements of the eyes, mouth, hands and other parts of the body. Whole-brain studies tell only part of the story; to understand the details of how signals pass to and from the prefrontal cortex, one must scrutinize the brain on a cellular scale.

When viewed through a conventional microscope, the cerebral cortex appears to be divided into six layers of varying cellular composition and density. Cells in each layer form their own set of connections within the brain. One class of cell, which resides in the fifth layer of the cortex, projects to areas beyond the cortex, including the caudate nucleus and putamen (which regulate a variety of motor activities) and the superior colliculus (which specifically processes visual motor functions). Neurons in the sixth layer of the cerebral cortex project into the thalamus, through which sensory inputs from the brain's periphery travel to reach the cortex (see Figure 6.5).

The prefrontal cortex probably cannot independently trigger motor responses. Nevertheless, it may regulate motor behavior by initiating, programming, facilitating and canceling commands to brain structures that are more immediately involved in directing muscular movement. Such commands are transmitted via an elaborate set of chemical pathways in the brain. Neuroscientists and biochemists around the world have been racing to learn more about these chemicals and how they regulate the operation of the brain.

A number of researchers studying rodent brains, including Anne Marie Thierry and Jacques Glowinski of the College of France in Paris, Brigitte Berger of Pitié Salpêtrière Hospital, also in Paris, and Tomas Hökfelt of the Karolinska Institute in Sweden, along with many colleagues, find that the prefrontal cortex abounds in catecholamines, a family of compounds that prepare the body for a stressful situation. Those compounds also act as neurotransmitters, substances that transmit neuronal impulses in the brain. My co-workers and I have discovered a similar abundance of catecholamines in the prefrontal cortices of nonhuman primates. One of the most familiar catecholamines, dopamine, regulates how neurons react to stimuli and seems to play a central role in schizophrenia.

A growing body of evidence suggests that dopamine is one of the most important of the chemicals that regulate cell activity associated with working memory. An imbalance in the abundance of dopamine in the prefrontal cortex can induce deficits in the working memory similar to those resulting from lesions in the principal sulcus region of the prefrontal cortex. For example, aged monkeys whose prefrontal cortices are deficient in dopamine and norepinephrine (a chemical relative of adrenaline) perform poorly in delayed-response tests. Injecting the aged animals with the deficient neurotransmitters restored their memory function so that they tested roughly as well as younger, healthy monkeys.

Many of my colleagues and I are striving to learn which cells respond to dopamine and how they affect working memory. Within the past several years, we have collected evidence showing that neurons in certain layers of the cerebral cortex contain a great abundance of D_1 receptors, one of the chemical sites where dopamine binds to a cell. Interestingly, the neurons that are rich in D_1 receptors are those that project to the thalamus, the brain structure that relays information to the cortex.

Csaba Leranth, John Smiley and F. Mark Williams of Yale are examining the cellular structures that enable dopamine to modulate responses to sensory inputs in the cerebral cortex. The researchers use an antibody developed by Michel Geffard of the Institute of Cellular Biochemistry and of Neurochemistry of the National Center of Scientific Research in Bordeaux, France, to label the neurons and their axonal projections that contain dopamine. They then scrutinize those cells under an electron microscope. The team looked in particular at the points of contact between dopamine-releasing cells and the neuronal spines, small protuberances where the cells receive incoming signals. Spines are discrete sites where calcium ions can enter and activate cellular mechanisms involving information processing and modulation of neuronal responses.

In most cases, the dopamine-releasing cells make symmetric contact with the spines—that is, the cell projections on either side of the synaptic cleft show roughly the same density. Such symmetric contacts are thought to have an inhibitory effect: when the postsynaptic site is activated, the cell's normal, spontaneous electrical activity is dampened. A large proportion of the spines of pyramidal cells—the major class of neuron that projects out of the cortex—receive asymmetric contacts from the axons of another cell whose point of origin has not yet been identified but which is thought to carry signals from other cortical areas. Those asymmetric contacts probably have an opposite, excitatory effect.

Pyramidal cells receive the major sensory or informational signals arriving at the cerebral cortex (see Figure 6.6). The network of excitatory and inhibitory synapses, or connections, noted by the Yale group provides a mechanism by which dopamine could alter the way that various classes of pyramidal neurons respond to integrate such signals across thousands of spines in their dendrites. In this way, dopamine may regulate the overall output of the cortex. Further analysis of the physical and chemical interactions between pyramidal cells and other neurons in the cerebral cortex should clarify how

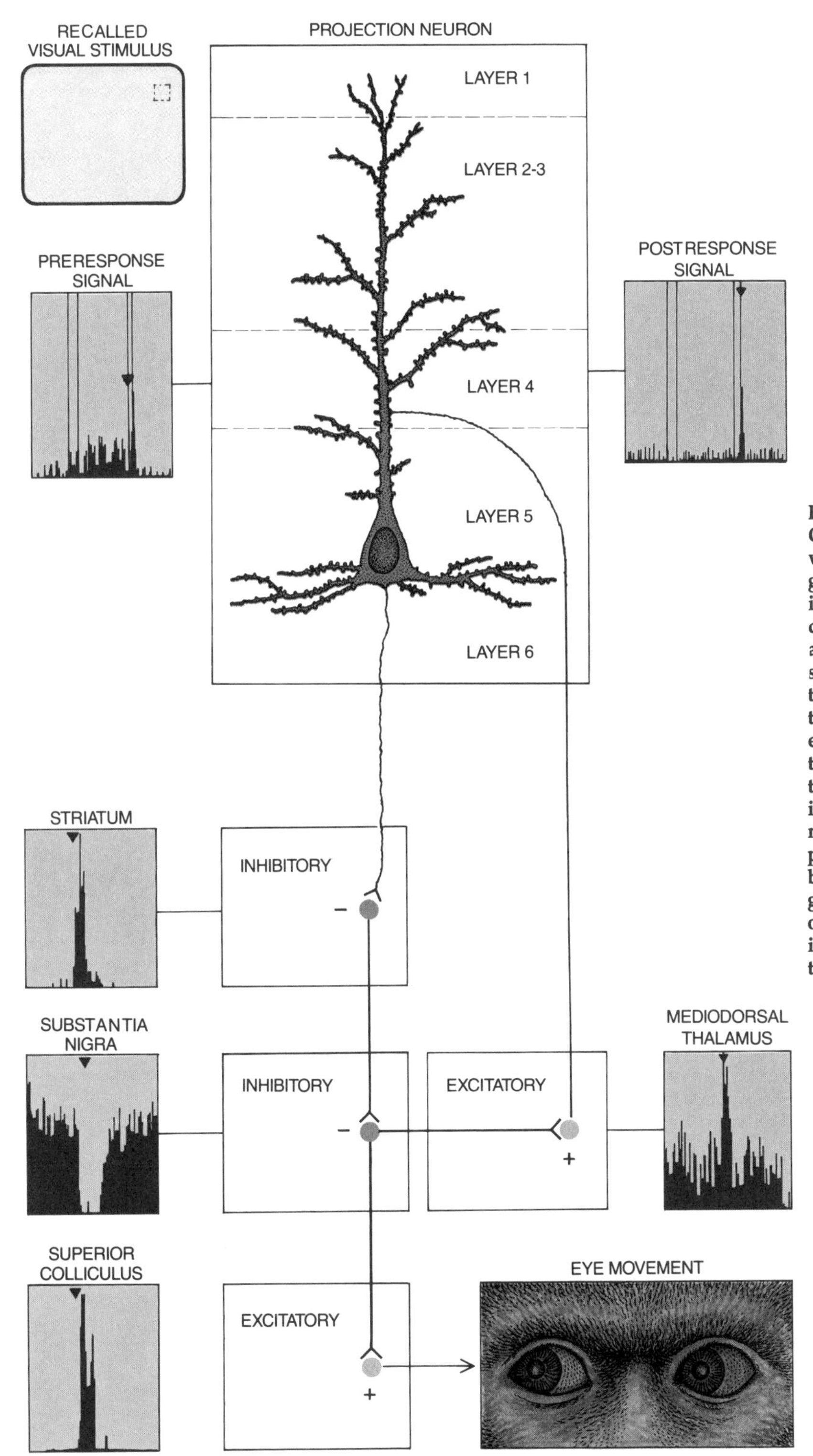

Figure 6.5 ELABORATE FLOW OF NEURAL SIGNALS is involved in producing a memory-guided eye movement. A neuron in the fifth layer of the prefrontal cerebral cortex transmits signals along a chain of neurons in the striatum, the substantia nigra and the superior colliculus, where they trigger motor response in the eyes. Impulses from the substantia nigra travel to the mediodorsal thalamus and back to the cortex, indicating the completion of the motor response and signaling the prefrontal neuron to return to a baseline level of activity. The graphs show the electrical activity of the neurons; inverted triangles indicate the nearly instantaneous travel of the signals.

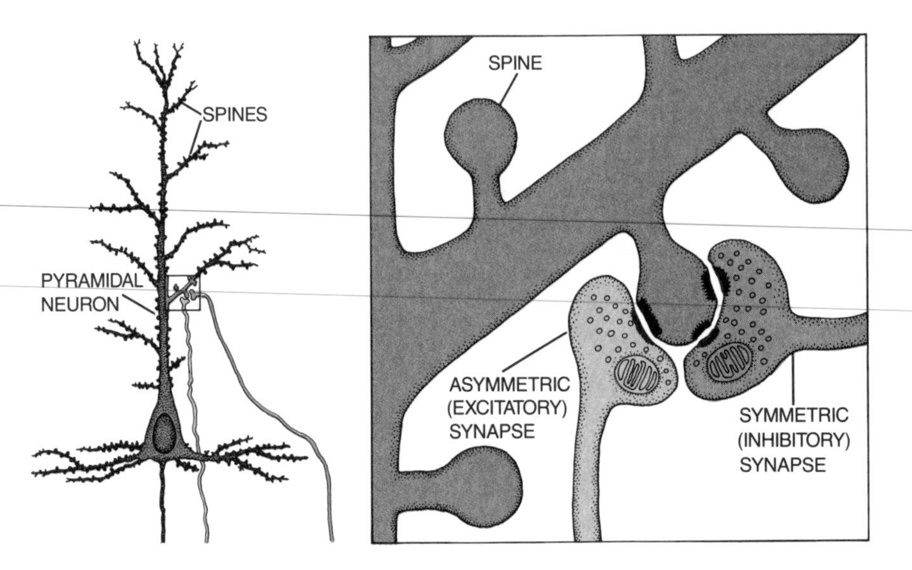

Figure 6.6 PYRAMIDAL NEURON (*left*) in the prefrontal cortex is thought to modulate signals to and from the prefrontal cortex. Each such neuron is covered with thousands of spines. bulblike projections where synaptic connections occur. Synapses have different morphologies depending on whether they are excitatory or inhibitory (*right*). The dopamine-containing connections in the cortex are of the inhibitory type.

dopamine and other neurotransmitters influence cognition by stimulating or repressing the cellular responses of cortical neurons.

Investigations of the workings of the prefrontal cortex are revealing not only how the mind operates but also what goes wrong when it malfunctions. Medical researchers have implicated dysfunction of the prefrontal cortex as the cause of many neurological and psychiatric disorders, including Parkinson's disease and especially schizophrenia. The abnormal mental attributes associated with schizophrenia strongly resemble those caused by physical damage to the prefrontal cortex: thought disorders, reduced attention span, inappropriate or flattened emotional responses and lack of initiative, plans and goals. Schizophrenic patients, like frontal lobe patients and monkeys afflicted with prefrontal lobe lesions, retain a normal ability to perform routine procedures or habits but exhibit fragmented, disorganized behavior when attempting to perform tasks involving symbolic or verbal information (see Chapter 8, "Major Disorders of Mind and Brain," by Elliot S. Gershon and Ronald O. Rieder).

Schizophrenic patients taking tests such as the Wisconsin Card Sort test tend to repeat a previous response even when it is clear that it is no longer the correct one; normal subjects, in contrast, shift hypotheses much sooner after making an error. Schizophrenic individuals are also severely impaired both on spatial delayed-response tasks and on a variety of tests of problem solving, abstraction and planning.

Studies of cerebral blood flow by David H. Ingvar of University Hospital in Lund, Sweden, and by Daniel R. Weinberger, Karen F. Berman and others at the National Institute of Mental Health, as well as measurements of local cerebral metabolism made by Monte S. Buchsbaum of the University of California at Irvine, show that schizophrenic patients have below-average blood flow into their prefrontal cortices, indicative of a depressed level of activity in that part of the brain. Schizophrenic subjects often suffer from impaired ability to move their eyes to track and project the forward trajectories of moving targets, further evidence that the disorder involves malfunctions in a posterior part of the prefrontal cortex, where the eye-movement centers involved in predictive tracking are located.

Sohee Park and Philip S. Holzman of Harvard University have shown that schizophrenic subjects exhibit impaired performance on working memory tasks much like those my colleagues and I have used to study working memory in rhesus monkeys. Conversely, Martha MacAvoy and Bruce of Yale have demonstrated that monkeys with lesions in the relevant portions of the prefrontal cortex exhibit the same type of predictive tracking disorder that has long been considered a marker of schizophrenia in humans.

Perhaps researchers should begin to think of schizophrenia as a breakdown in the processes by which representational knowledge governs behavior. In my view, neural pathways in the prefrontal cortex update inner models of reality to reflect changing environmental demands and incoming information. Those pathways guide short-term memory and moment-to-moment behavior. If they fail, the brain views the world as a series of disconnected events, like a slide show, rather than as a continuous sequence, like a movie. The result is schizophrenic behavior, excessively dominated by immediate stimulation rather than by a balance of current, internal and past information.

At present, theories describing the fundamental causes of schizophrenia are inadequate, much as knowledge of the functioning of the working memory system remains frustratingly sketchy. Fortunately, neurobiological research has been advancing at a breathless pace in the past few years. Such research should lead to a greater understanding not only of schizophrenia but of the prefrontal cortex and how it shapes short-term memory and the broader working of the rational mind.

Sex Differences in the Brain

Cognitive variations between the sexes reflect differing hormonal influences on brain development. Understanding these differences and their causes can yield insights into brain organization.

• • •

Doreen Kimura

Women and men differ not only in physical attributes and reproductive function but also in the way in which they solve intellectual problems. It has been fashionable to insist that these differences are minimal, the consequence of variations in experience during development. The bulk of the evidence suggests, however, that the effects of sex hormones on brain organization occur so early in life that from the start the environment is acting on differently wired brains in girls and boys. Such differences make it almost impossible to evaluate the effects of experience independent of physiological predisposition.

Behavioral, neurological and endocrinologic studies have elucidated the processes giving rise to sex differences in the brain. As a result, aspects of the physiological basis for these variations have in recent years become clearer. In addition, studies of the effects of hormones on brain function throughout life suggest that the evolutionary pressures directing differences nevertheless allow for a degree of flexibility in cognitive ability between the sexes.

Figure 7.1 ROUTES in a landscape such as the one in this painting, *The Old Oaken Bucket*, by Grandma Moses (1860–1961), may be learned differently by women and men. In laboratory experiments, researchers have found that women tend to remember landmarks—such as the well at the lower right or a tree at an intersection. Men appear to learn routes faster, but they cannot recall landmarks as readily: they may rely preferentially on spatial cues such as distance and direction.

Major sex differences in intellectual function seem to lie in patterns of ability rather than in overall level of intelligence (IQ). We are all aware that people have different intellectual strengths. Some are especially good with words, others at using objects—for instance, at constructing or fixing things. In the same fashion, two individuals may have the same overall intelligence but have varying patterns of ability.

Men, on average, perform better than women on certain spatial tasks. In particular, men have an advantage in tests that require the subject to imagine rotating an object or manipulating it in some other way. They outperform women in mathematical reasoning tests and in navigating their way through a route. Further, men are more accurate in tests of target-directed motor skills—that is, in guiding or intercepting projectiles.

Women tend to be better than men at rapidly identifying matching items, a skill called perceptual speed. They have greater verbal fluency, including the ability to find words that begin with a specific letter or fulfill some other constraint. Women also outperform men in arithmetic calculation and in recalling landmarks from a route. Moreover, women are faster at certain precision manual tasks, such as placing pegs in designated holes on a board.

Although some investigators have reported that sex differences in problem solving do not appear until after puberty, Diane Lunn, working in my laboratory at the University of Western Ontario, and I have found three-year-old boys to be better at targeting than girls of the same age. Moreover, Neil V. Watson, when in my laboratory, showed that the extent of experience playing sports does not account for the sex difference in targeting found in young adults. Kimberly A. Kerns, working with Sheri A. Berenbaum of the University of Chicago has found that sex differences in spatial rotation performance are present before puberty.

Differences in route learning have been systematically studied in adults in laboratory situations (see Figure 7.1). For instance, Liisa Galea in my department studied undergraduates who followed a route on a tabletop map. Men learned the route in fewer trials and made fewer errors than did women. But once learning was complete, women remembered more of the landmarks than did men. These results, and those of other researchers, raise the possibility that women tend to use landmarks as a strategy to orient themselves in everyday life. The prevailing strategies used by males have not yet been clearly established, although they must relate to spatial ability.

Marion Eals and Irwin Silverman of York University studied another function that may be related to landmark memory. The researchers tested the ability of individuals to recall objects and their locations within a confined space—such as in a room or on a tabletop. Women were better able to remember whether an item had been displaced or not. In addition, in my laboratory, we measured the accuracy of object location: subjects were shown an array of objects and were later asked to replace them in their exact positions. Women did so more accurately than

Problem-Solving Tasks Favoring Women

Women tend to perform better than men on tests of perceptual speed, in which subjects must rapidly identify matching items—for example, pairing the house on the far left with its twin:

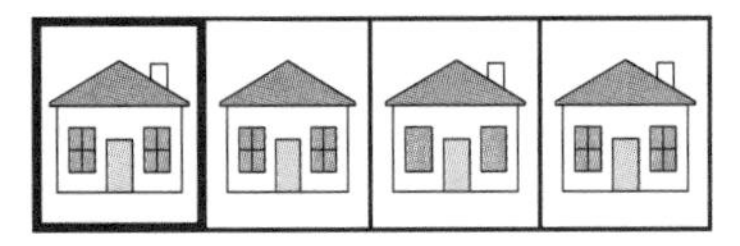

In addition, women remember whether an object, or a series of objects, has been displaced:

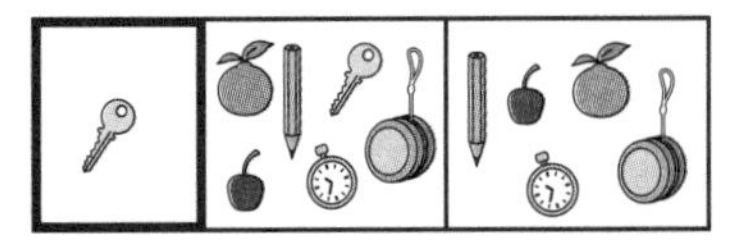

On some tests of ideational fluency, for example, those in which subjects must list objects that are the same color, and on tests of verbal fluency, in which participants must list words that begin with the same letter, women also outperform men:

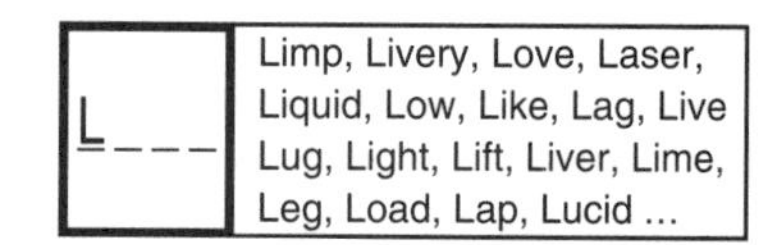

Women do better on precision manual tasks—that is, those involving fine-motor coordination—such as placing the pegs in holes on a board:

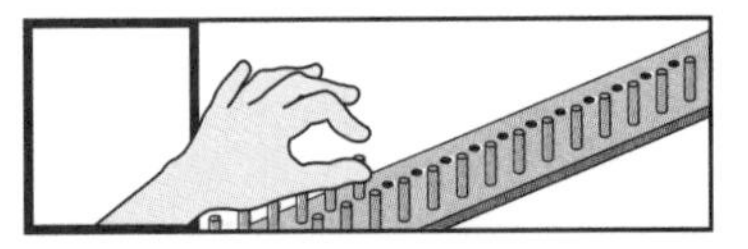

And women do better than men on mathematical calculation tests:

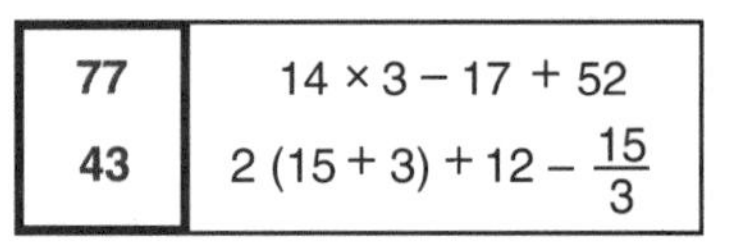

Problem-Solving Tasks Favoring Men

Men tend to perform better than women on certain spatial tasks. They do well on tests that involve mentally rotating an object or manipulating it in some fashion, such as imagining turning this three-dimensional object:

or determining where the holes punched in a folded piece of paper will fall when the paper is unfolded:

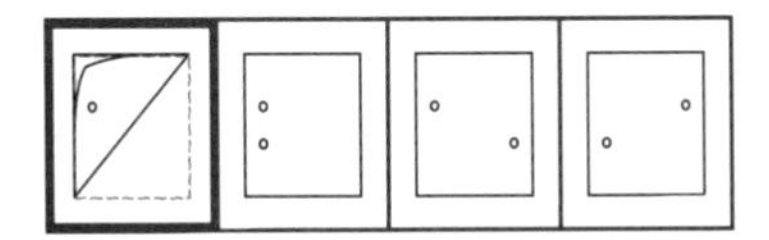

Men also are more accurate than women in target-directed motor skills, such as guiding or intercepting projectiles:

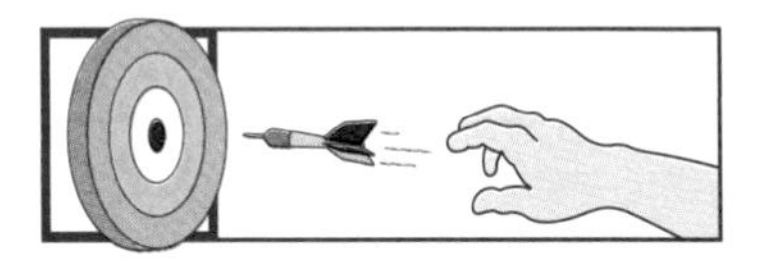

They do better on disembedding tests, in which they have to find a simple shape, such as the one on the left, once it is hidden within a more complex figure:

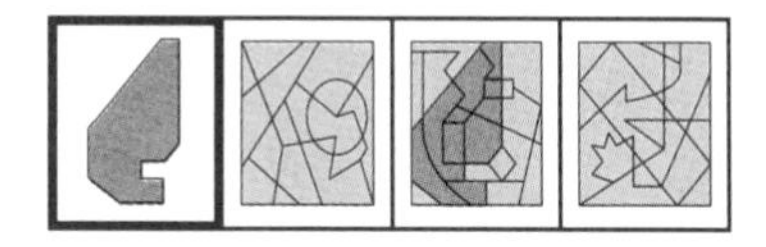

And men tend to do better than women on tests of mathematical reasoning:

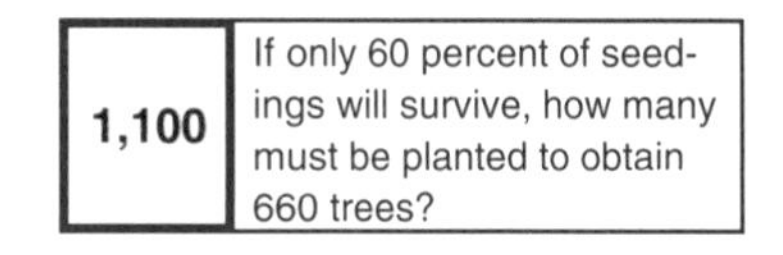

did men (see boxed figures "Problem-Solving Tasks Favoring Women" and "Problem-Solving Tasks Favoring Men").

It is important to place the differences described above in context: some are slight, some are quite large. Because men and women overlap enormously on many cognitive tests that show average sex differences, researchers use variations within each group as a tool to gauge the differences between groups. Imagine, for instance, that on one test the average score is 105 for women and 100 for men. If the scores for women ranged from 100 to 110 and for men from 95 to 105, the difference would be more impressive than if the women's scores ranged from 50 to 150 and the men's from 45 to 145. In the latter case, the overlap in scores would be much greater.

One measure of the variation of scores within a group is the standard deviation. To compare the magnitude of a sex difference across several distinct tasks, the difference between groups is divided by the standard deviation. The resulting number is called the effect size. Effect sizes below 0.5 are generally considered small. Based on my data, for instance, there are typically no differences between the sexes on tests of vocabulary (effect size 0.02), nonverbal reasoning (0.03) and verbal reasoning (0.17).

On tests in which subjects match pictures, find words that begin with similar letters or show ideational fluency—such as naming objects that are white or red—the effect sizes are somewhat larger: 0.25, 0.22 and 0.38, respectively. As discussed above, women tend to outperform men on these tasks. Researchers have reported the largest effect sizes for certain tests measuring spatial rotation (effect size 0.7) and targeting accuracy (0.75). The large effect size in these tests means there are many more men at the high end of the score distribution.

Since, with the exception of the sex chromosomes, men and women share genetic material, how do such differences come about? Differing patterns of ability between men and women most probably reflect different hormonal influences on their developing brains. Early in life the action of estrogens and androgens (male hormones chief of which is testosterone) establishes sexual differentiation. In mammals, including humans, the organism has the potential to be male or female. If a Y chromosome is present, testes or male gonads form. This development is the critical first step toward becoming a male. If the gonads do not produce male hormones or if for some reason the hormones cannot act on the tissue, the default form of the organism is female.

Once testes are formed, they produce two substances that bring about the development of a male. Testosterone causes masculinization by promoting the male, or Wolffian, set of ducts and, indirectly through conversion to dihydrotestosterone, the external appearance of scrotum and penis. The Müllerian regression factor causes the female, or Müllerian, set of ducts to regress. If anything goes wrong at any stage of the process, the individual may be incompletely masculinized.

Not only do sex hormones achieve the transformation of the genitals into male organs, but they also organize corresponding male behaviors early in life. Since we cannot manipulate the hormonal environment in humans, we owe much of what we know about the details of behavioral determination to studies in other animals. Again, the intrinsic tendency, according to studies by Robert W. Goy of the University of Wisconsin, is to develop the female pattern that occurs in the absence of masculinizing hormonal influence.

If a rodent with functional male genitals is deprived of androgens immediately after birth (either by castration or by the administration of a compound that blocks androgens), male sexual behavior, such as mounting, will be reduced. Instead female sexual behavior, such as lordosis (arching of the back), will be enhanced in adulthood. Similarly, if androgens are administered to a female directly after birth, she displays more male sexual behavior and less female behavior in adulthood.

Bruce S. McEwen and his co-workers at the Rockefeller University have shown that, in the rat, the two processes of defeminization and masculinization require somewhat different biochemical changes. These events also occur at somewhat different times. Testosterone can be converted to either estrogen (usually considered a female hormone) or dihydrotestosterone. Defeminization takes place primarily after birth in rats and is mediated by estrogen, whereas masculinization involves both dihydrotestosterone and estrogen and occurs for the most part before birth rather than after, according to studies by McEwen. A substance called alphafetoprotein may protect female brains from the masculinizing effects of their estrogen.

The area in the brain that organizes female and male reproductive behavior is the hypothalamus. This tiny structure at the base of the brain connects to the pituitary, the master endocrine gland. Roger A. Gorski and his colleagues at the University of California at Los Angeles have shown that a region of the pre-optic area of the hypothalamus is visibly larger in male rats than in females. The size increment in males is promoted by the presence of androgens in the immediate postnatal, and to some extent prenatal, period. Laura S. Allen in Gorski's laboratory has found a similar sex difference in the human brain.

Other preliminary but intriguing studies suggest that sexual behavior may reflect further anatomic differences. In 1991 Simon LeVay of the Salk Institute for Biological Studies in San Diego reported that one of the brain regions that is usually larger in human males than in females—an interstitial nucleus of the anterior hypothalamus—is smaller in homosexual than in heterosexual men. LeVay points out that this finding supports suggestions that sexual preference has a biological substrate.

Homosexual and heterosexual men may also perform differently on cognitive tests. Brian A. Gladue of North Dakota State University and Geoff D. Sanders of City of London Polytechnic report that homosexual men perform less well on several spatial tasks than do heterosexual men. In a recent study in my laboratory, Jeff Hall found that homosexual men had lower scores on targeting tasks than did heterosexual men; however, they were superior in ideational fluency—listing things that were a particular color.

This exciting field of research is just starting, and it is crucial that investigators consider the degree to which differences in life-style contribute to group differences. One should also keep in mind that results concerning group differences constitute a general statistical statement; they establish a mean from which any individual may differ. Such studies are potentially a rich source of information on the physiological basis for cognitive patterns.

The lifelong effects of early exposure to sex hormones are characterized as organizational, because they appear to alter brain function permanently during a critical period. Administering the same hormones at later stages has no such effect. The hormonal effects are not limited to sexual or reproductive behaviors: they appear to extend to all known behaviors in which males and females differ. They seem to govern problem solving, aggression and the tendency to engage in rough-and-tumble play—the boisterous body contact that young males of some mammalian species display. For example, Michael J. Meaney of McGill University finds that dihydrotestosterone, working through a structure called the amygdala rather than through the hypothalamus, gives rise to the play-fighting behavior of juvenile male rodents.

Male and female rats have also been found to solve problems differently. Christina L. Williams of Barnard College has shown that female rats have a greater tendency to use landmarks in spatial learning tasks—as it appears women do. In Williams's experiment, female rats used landmark cues, such as pictures on the wall, in preference to geometric cues, such as angles and the shape of the room. If no landmarks were available, however, females used geometric cues. In contrast, males did not use landmarks at all, preferring geometric cues almost exclusively.

Interestingly, hormonal manipulation during the critical period can alter these behaviors. Depriving newborn males of testosterone by castrating them or administering estrogen to newborn females results in a complete reversal of sex-typed behaviors in the adult animals. (As mentioned above, estrogen can have a masculinizing effect during brain development.) Treated females behave like males, and treated males behave like females.

Natural selection for reproductive advantage could account for the evolution of such navigational differences. Steven J. C. Gaulin and Randall W. FitzGerald of the University of Pittsburgh have suggested that in species of voles in which a male mates with several females, rather than with just one, the range he must traverse is greater. Therefore, navigational ability seems critical to reproductive success. Indeed, Gaulin and FitzGerald found sex differences in laboratory maze learning only in voles that were polygynous, such as the meadow vole, not in monogamous species, such as the prairie vole.

Again, behavioral differences may parallel structural ones. Lucia F. Jacobs in Gaulin's laboratory has discovered that the hippocampus—a region thought to be involved in spatial learning in both birds and mammals—is larger in male polygynous voles than in females. At present, there are no data on possible sex differences in hippocampal size in human subjects.

Evidence of the influence of sex hormones on adult behavior is less direct in humans than in other animals. Researchers are instead guided by what may be parallels in other species and by spontaneously occurring exceptions to the norm in humans.

One of the most compelling areas of evidence comes from studies of girls exposed to excess androgens in the prenatal or neonatal stage. The production of abnormally large quantities of adrenal androgens can occur because of a genetic defect called congenital adrenal hyperplasia (CAH). Before the 1970s, a similar condition also unexpectedly appeared when pregnant women took various synthetic steroids. Although the consequent masculinization of the genitals can be corrected early in life, and drug therapy can stop the overproduction of androgens, effects of prenatal exposure on the brain cannot be reversed.

Studies by researchers such as Anke A. Ehrhardt of Columbia University and June M. Reinisch of the Kinsey Institute have found that girls with excess exposure to androgens grow up to be more tomboyish and aggressive than their unaffected sisters. This conclusion was based sometimes on interviews with subjects and mothers, on teachers' ratings and on questionnaires administered to the girls themselves. When ratings are used in such studies, it can be difficult to rule out the influence of expectation either on the part of an adult who knows the girls' history or on the part of the girls themselves.

Therefore, the objective observations of Berenbaum are important and convincing. She and Melissa Hines of the University of California at Los Angeles observed the play behavior of CAH-affected girls and compared it with that of their male and female siblings. Given a choice of transportation and construction toys, dolls and kitchen supplies or books and board games, the CAH girls preferred the more typically masculine toys—for example, they played with cars for the same amount of time that normal boys did. Both the CAH girls and the boys differed from unaffected girls in their patterns of choice. Because there is every reason to think that parents would be at least as likely to encourage feminine preferences in their CAH daughters as in their unaffected daughters, these findings suggest that the toy preferences were actually altered in some way by the early hormonal environment.

Spatial abilities that are typically better in males

are also enhanced in CAH girls. Susan M. Resnick, now at the National Institute on Aging, and Berenbaum and their colleagues reported that affected girls were superior to their unaffected sisters in a spatial manipulation test, two spatial rotation tests and a disembedding test—that is, the discovery of a simple figure hidden within a more complex one. All these tasks are usually done better by males. No differences existed between the two groups on other perceptual or verbal tasks or on a reasoning task.

Studies such as these suggest that the higher the androgen levels, the better the spatial performance. But this does not seem to be the case. In 1983 Valerie J. Shute, when at the University of California at Santa Barbara, suggested that the relation between levels of androgens and some spatial capabilities might be nonlinear. In other words, spatial ability might not increase as the amount of androgen increases. Shute measured androgens in blood taken from male and female students and divided each into high- and low-androgen groups. All fell within the normal range for each sex (androgens are present in females but in very low levels). She found that in women, the high-androgen subjects were better at the spatial tests. In men the reverse was true: low-androgen men performed better.

Catherine Gouchie and I recently conducted a study along similar lines by measuring testosterone in saliva. We added tests for two other kinds of abilities: mathematical reasoning and perceptual speed. Our results on the spatial tests were very similar to Shute's: low-testosterone men were superior to high-testosterone men, but high-testosterone women surpassed low-testosterone women. Such findings suggest some optimum level of androgen for maximal spatial ability. This level may fall in the low male range.

No correlation was found between testosterone levels and performance on perceptual speed tests. On mathematical reasoning, however, the results were similar to those of spatial ability tests for men: low-androgen men tested higher, but there was no obvious relation in women (see Figure 7.2).

Such findings are consistent with the suggestion by Camilla P. Benbow of Iowa State University that high mathematical ability has a significant biological determinant. Benbow and her colleagues have reported consistent sex differences in mathematical reasoning ability favoring males. These differences are especially sharp at the upper end of the distribution, where males outnumber females 13 to one. Benbow argues that these differences are not readily explained by socialization.

It is important to keep in mind that the relation between natural hormonal levels and problem solving is based on correlational data. Some form of connection between the two measures exists, but how this association is determined or what its causal basis may be is unknown. Little is currently understood about the relation between adult levels of hormones and those in early life, when abilities appear to be organized in the nervous system. We have a lot to learn about the precise mechanisms underlying cognitive patterns in people.

Another approach to probing differences between male and female brains is to examine and compare the functions of particular brain systems. One noninvasive way to accomplish this goal is to study people who have experienced damage to a specific brain region. Such studies indicate that the left half of the brain in most people is critical for speech, the right for certain perceptual and spatial functions.

It is widely assumed by many researchers studying sex differences that the two hemispheres are more asymmetrically organized for speech and spatial functions in men than in women. This idea comes from several sources. Parts of the corpus callosum, a major neural system connecting the two hemispheres, may be more extensive in women; perceptual techniques that probe brain asymmetry in normal-functioning people sometimes show smaller asymmetries in women than in men, and damage to one brain hemisphere sometimes has a lesser effect in women than the comparable injury has in men.

In 1982 Marie-Christine de Lacoste, now at the Yale University School of Medicine, and Ralph L. Holloway of Columbia University reported that the back part of the corpus callosum, an area called the splenium, was larger in women than in men. This finding has subsequently been both refuted and confirmed. Variations in the shape of the corpus callosum that may occur as an individual ages as well as different methods of measurement may produce some of the disagreements. Most recently, Allen and Gorski found the same sex-related size difference in the splenium.

The interest in the corpus callosum arises from the assumption that its size may indicate the number of fibers connecting the two hemispheres. If more connecting fibers existed in one sex, the implication would be that in that sex the hemispheres

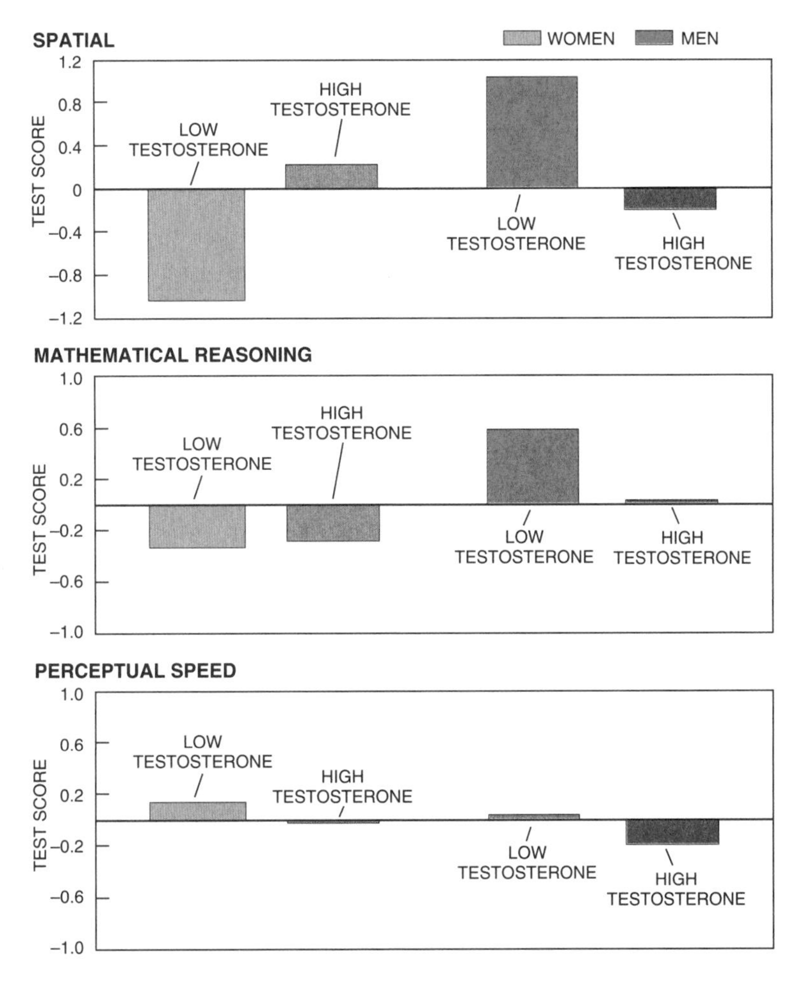

Figure 7.2 TESTOSTERONE LEVELS can affect performance on some tests. Women with high levels of testosterone perform better on a spatial task (*top*) than do women with low levels; men with low levels outperform men with high levels. On a mathematical reasoning test (*middle*), low testosterone corresponds to better performance in men; in women there is no such relation. On a test in which women usually excel (*bottom*), no relation is found between testosterone and performance.

communicate more fully. Although sex hormones can alter callosal size in rats, as Victor H. Denenberg and his associates at the University of Connecticut have demonstrated, it is unclear whether the actual number of fibers differs between the sexes. Moreover, sex differences in cognitive function have yet to be related to a difference in callosal size. New ways of imaging the brain in living humans will undoubtedly increase knowledge in this respect.

The view that a male brain is functionally more asymmetric than a female brain is long-standing. Albert M. Galaburda of Beth Israel Hospital in Boston and the late Norman Geschwind of Harvard Medical School proposed that androgens increased the functional potency of the right hemisphere. In 1981 Marian C. Diamond of the University of California at Berkeley found that the right cortex is thicker than the left in male rats but not in females. Jane Stewart of Concordia University in Montreal, working with Bryan E. Kolb of the University of Lethbridge in Alberta, recently pinpointed early hormonal influences on this asymmetry: androgens appear to suppress left cortex growth.

Last year de Lacoste and her colleagues reported a similar pattern in human fetuses. They found the right cortex was thicker than the left in males. Thus,

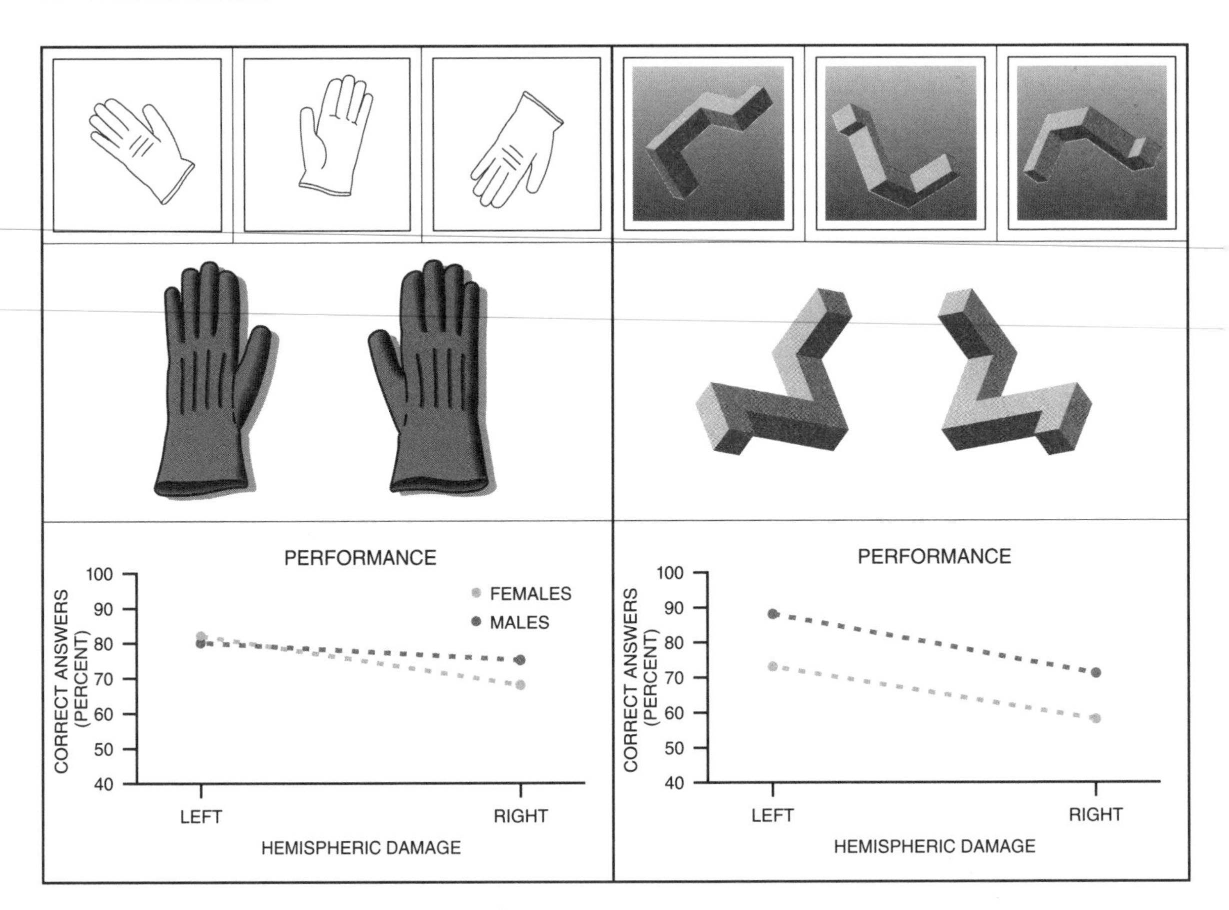

Figure 7.3 RIGHT HEMISPHERIC DAMAGE affects spatial ability to the same degree in both sexes (*graphs at bottom*), suggesting that women and men rely equally on that hemisphere for certain spatial tasks. In one test of spatial rotation performance (*left*), a series of drawings of a gloved right or left hand must be matched to a right- or left-handed glove. In a second test (*right*), photographs of a three-dimensional object must be matched to one of two mirror images of the same object.

there appear to be some anatomic reasons for believing that the two hemispheres might not be equally asymmetric in men and women.

Despite this expectation, the evidence in favor of it is meager and conflicting, which suggests that the most striking sex differences in brain organization may not be related to asymmetry. For example, if overall differences between men and women in spatial ability were related to differing right hemispheric dependence for such functions, then damage to the right hemisphere would perhaps have a more devastating effect on spatial performance in men.

My laboratory has recently studied the ability of patients with damage to one hemisphere of the brain to rotate certain objects mentally. In one test, a series of line drawings of either a left or a right gloved hand is presented in various orientations. The patient indicates the hand being depicted by simply pointing to one of two stuffed gloves that are constantly present.

The second test uses two three-dimensional blocklike figures that are mirror images of one another. Both figures are present throughout the test. The patient is given a series of photographs of these objects in various orientations, and he or she must place each picture in front of the object it depicts. (These nonverbal procedures are employed so that patients with speech disorders can be tested.)

As expected, damage to the right hemisphere resulted in lower scores for both sexes on these tests than did damage to the left hemisphere (see Figure

7.3). Also as anticipated, women did less well than men on the block spatial rotation test. Surprisingly, however, damage to the right hemisphere had no greater effect in men than in women. Women were at least as affected as men by damage to the right hemisphere. This result suggests that the normal differences between men and women on such rotational tests are not the result of differential dependence on the right hemisphere. Some other brain systems must be mediating the higher performance by men.

Parallel suggestions of greater asymmetry in men regarding speech have rested on the fact that the incidence of aphasias, or speech disorders, are higher in men than in women after damage to the left hemisphere. Therefore, some researchers have found it reasonable to conclude that speech must be more bilaterally organized in women. There is, however, a problem with this conclusion. During my 20 years of experience with patients, aphasia has not been disproportionately present in women with right hemispheric damage.

In searching for an explanation, I discovered another striking difference between men and women in brain organization for speech and related motor function. Women are more likely than men to suffer aphasia when the front part of the brain is damaged (see Figure 7.4). Because restricted damage within a hemisphere more frequently affects the posterior than the anterior area in both men and women, this differential dependence may explain why women incur aphasia less often than do men. Speech functions are thus less likely to be affected in women not because speech is more bilaterally organized in women but because the critical area is less often affected.

A similar pattern emerges in studies of the control of hand movements, which are programmed by the left hemisphere. Apraxia, or difficulty in selecting appropriate hand movements, is very common after left hemispheric damage. It is also strongly associated with difficulty in organizing speech (see Figure 7.5). In fact, the critical functions that depend on the left hemisphere may relate not to language

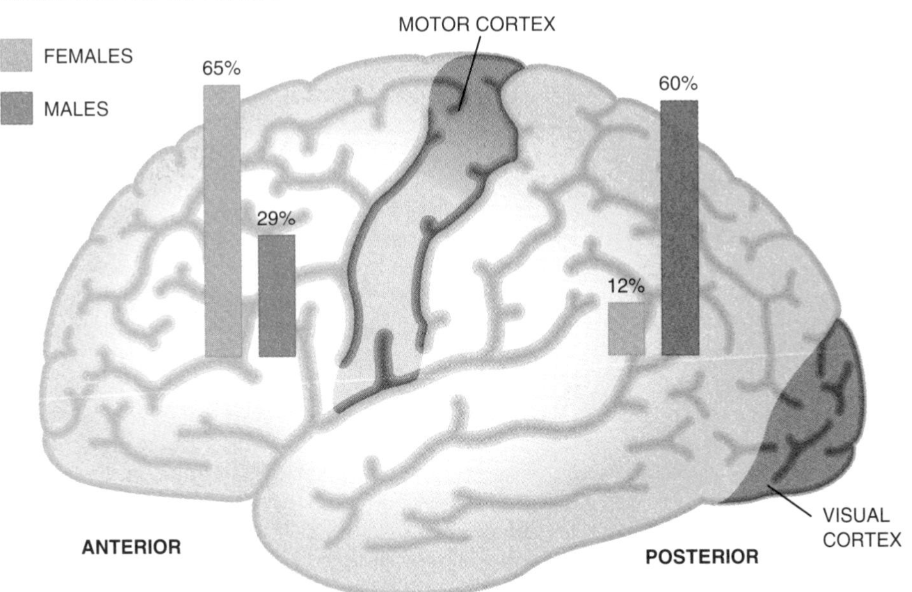

Figure 7.4 APHASIAS, or speech disorders, occur most often in women when damage is to the front of the brain. In men, they occur more frequently when damage is in the posterior region. The data presented above derive from one set of patients.

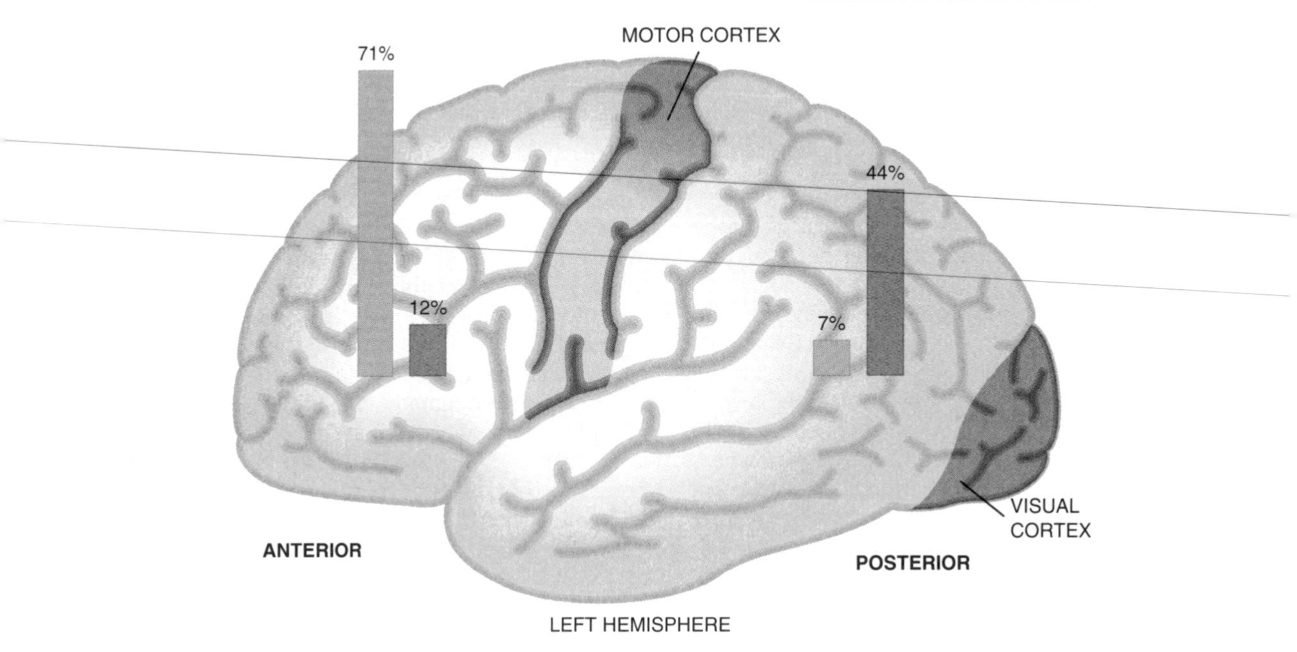

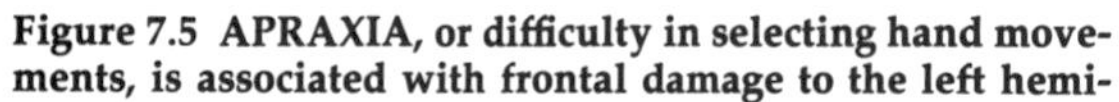

Figure 7.5 APRAXIA, or difficulty in selecting hand movements, is associated with frontal damage to the left hemisphere in women and with posterior damage in men. It is also associated with difficulties in organizing speech.

per se but to organization of the complex oral and manual movements on which human communication systems depend. Studies of patients with left hemispheric damage have revealed that such motor selection relies on anterior systems in women but on posterior systems in men.

The synaptic proximity of women's anterior motor selection system (or "praxis system") to the motor cortex directly behind it may enhance fine-motor skills. In contrast, men's motor skills appear to emphasize targeting or directing movements toward external space—some distance away from the self. There may be advantages to such motor skills when they are closely meshed with visual input to the brain, which lies in the posterior region.

Women's dependence on the anterior region is detectable even when tests involve using visual guidance—for instance, when subjects must build patterns with blocks by following a visual model. In studying such a complex task, it is possible to compare the effects of damage to the anterior and posterior regions of both hemispheres because performance is affected by damage to either hemisphere. Again, women prove more affected by damage to the anterior region of the right hemisphere than by posterior damage. Men tend to display the reverse pattern.

Although I have not found evidence of sex differences in functional brain asymmetry with regard to basic speech, motor selection or spatial rotation ability, I have found slight differences in more abstract verbal tasks. Scores on a vocabulary test, for instance, were affected by damage to either hemisphere in women, but such scores were affected only by left-sided injury in men. This finding suggests that in reviewing the meanings of words, women use the hemispheres more equally than do men.

In contrast, the incidence of non-right-handedness, which is presumably related to lesser left hemispheric dependence, is higher in men than in women. Even among right-handers, Marion Annett, now at the University of Leicester in the U.K., has reported that women are more right-handed than men—that is, they favor their right hand even more than do right-handed men. It may well be, then, that sex differences in asymmetry vary with the particular function being studied and that it is not always the same sex that is more asymmetric.

Taken altogether, the evidence suggests that

men's and women's brains are organized along different lines from very early in life. During development, sex hormones direct such differentiation. Similar mechanisms probably operate to produce variation within sexes, since there is a relation between levels of certain hormones and cognitive makeup in adulthood.

One of the most intriguing findings is that cognitive patterns may remain sensitive to hormonal fluctuations throughout life. Elizabeth Hampson of the University of Western Ontario showed that the performance of women on certain tasks changed throughout the menstrual cycle as levels of estrogen went up or down. High levels of the hormone were associated not only with relatively depressed spatial ability but also with enhanced articulatory and motor capability.

In addition, I have observed seasonal fluctuations in spatial ability in men. Their performance is improved in the spring when testosterone levels are lower. Whether these intellectual fluctuations are of any adaptive significance or merely represent ripples on a stable baseline remains to be determined.

To understand human intellectual functions, including how groups may differ in such functions, we need to look beyond the demands of modern life. We did not undergo natural selection for reading or for operating computers. It seems clear that the sex differences in cognitive patterns arose because they proved evolutionarily advantageous. And their adaptive significance probably rests in the distant past. The organization of the human brain was determined over many generations by natural selection. As studies of fossil skulls have shown, our brains are essentially like those of our ancestors of 50,000 or more years ago.

For the thousands of years during which our brain characteristics evolved, humans lived in relatively small groups of hunter-gatherers. The division of labor between the sexes in such a society probably was quite marked, as it is in existing hunter-gatherer societies. Men were responsible for hunting large game, which often required long-distance travel. They were also responsible for defending the group against predators and enemies and for the shaping and use of weapons. Women most probably gathered food near the camp, tended the home, prepared food and clothing and cared for children.

Such specializations would put different selection pressures on men and women. Men would require long-distance route-finding ability so they could recognize a geographic array from varying orientations. They would also need targeting skills. Women would require short-range navigation, perhaps using landmarks, fine-motor capabilities carried on within a circumscribed space, and perceptual discrimination sensitive to small changes in the environment or in children's appearance or behavior.

The finding of consistent and, in some cases, quite substantial sex differences suggests that men and women may have different occupational interests and capabilities, independent of societal influences. I would not expect, for example, that men and women would necessarily be equally represented in activities or professions that emphasize spatial or math skills, such as engineering or physics. But I might expect more women in medical diagnostic fields where perceptual skills are important. So that even though any one individual might have the capacity to be in a "nontypical" field, the sex proportions as a whole may vary.

Major Disorders of Mind and Brain

Schizophrenia and manic-depressive illness are shaped by heredity and marked by structural and biochemical changes in the brain. The predisposing genes remain unknown.

• • •

Elliot S. Gershon and Ronald O. Rieder

Madness was understood for centuries by religion and poetry as an affliction of the spirit and by medicine as a disorder of various humors and organs of the body. In the past century, physicians have recognized the most common forms of psychosis (our current word for madness) as two chronic disorders—schizophrenia and mania—and have begun to understand the abnormalities in brain structure and function that accompany them. Each affects about 1 percent of the population. Both flare episodically, although schizophrenia follows a deteriorating course, whereas patients with bipolar manic-depressive illness, who have episodes of mania and depression, are usually mentally normal between episodes.

The anatomic, biochemical and hereditary bases of these disorders are now emerging. Some research has already shaped the development of new treatments. These subjects form the focus of this chapter. First, however, it is useful to consider what these disorders are like for the people who have them.

Figure 8.1 HALLUCINATORY SELF-PORTRAIT exemplifies the visual creativity that schizophrenia sometimes induces; it also conveys the pain and perceptional distortions of this illness.

When Mrs. T. was 16 years old, she began to experience her first symptom of schizophrenia: a profound feeling that people were staring at her. These bouts of self-consciousness soon forced her to end her public piano performances. Her self-consciousness led to withdrawal, then to fearful delusions that others were speaking of her and finally to suspicions that they were plotting to harm her. At first Mrs. T.'s illness was intermittent, and the return of her intelligence, warmth and ambition between episodes allowed her to complete several years of college, to marry and to rear three children. She had to enter a hospital for the first time at 28, after the birth of her third child, when she began to hallucinate.

Now, at 45, Mrs. T. is never entirely well. She has seen dinosaurs on the street and live animals in her refrigerator. While hallucinating, she speaks and writes in an incoherent, but almost poetic, way. At other times, she is more lucid, but even then her voices sometimes lead her to do dangerous things, such as driving very fast down the highway in the middle of the night, dressed only in a nightgown. As an episode winds down, Mrs. T. usually becomes deeply depressed and hopeless about her condition. Often she sits in her car with the engine running and contemplates committing suicide.

Over the past five years she has taken antipsychotic medications, such as haloperidol, that suppress the hallucinations and help her stay out of the hospital. Stress, however, can bring the hallucinations and delusions back for days or weeks, as happened after her recent separation from her husband and the subsequent sale of her home. At such times, her voices shout terrible criticisms. After her daughter left for college, they shouted, "You'll never see her again, you have been a bad mother, she'll die." At other times and without any apparent stimulus, Mrs. T. has bizarre visual hallucinations. For example, she saw cherubs in the grocery store. These experiences leave her preoccupied, confused and frightened, unable to perform such everyday tasks as cooking or playing the piano. When feeling well, however, she does volunteer work at church.

The mood disorders, which are distinct from schizophrenia, are called unipolar when the patient has episodes of depression alone and bipolar when there are episodes of both mania and depression. (The term "manic-depressive illness" encompasses both the unipolar and the bipolar form; the term "bipolar" is also used for the rare cases in which mania occurs without depression.) The depressions are quite severe, and suicide is an all too frequent outcome. Mania, a state of excitement usually characterized by impulsive behavior, can, when untreated, ultimately ruin marriages, careers and fortunes.

Mania can develop suddenly and shockingly, as illustrated in a case cited by a group led by Robert L. Spitzer of Columbia University. Daryl, a 25-year-old dancer, was cast in a part for a Broadway show. He began to come home at the end of rehearsal making disparaging remarks about the sessions and the director. A week later a fellow performer called Daryl's wife to complain that her husband had been trying to take over the rehearsals, giving unsolicited advice to the director and the other performers. At this point, his wife realized that Daryl's usually easygoing demeanor had turned tense and irritable. He began making nasty comments about his wife's figure and their recent sex life. Three days later he began shouting obscenities at other performers and was ejected from the theater. At home, he talked "a mile a minute" and paced incessantly, dressed only in his underwear. He felt no need to eat or sleep. The next day he skipped work to make a number of extravagant purchases.

At this time, two weeks after he had displayed his first symptoms, Daryl accepted hospitalization. He received one dose of a tranquilizer yet spent most of the night disrupting the ward. Then he signed out, against medical advice, in the morning. Eventually he responded well to lithium carbonate. Daryl's father has had a similar but more prolonged history, losing many jobs over 20 years after episodes of excited confrontations with his bosses. Over the past five years, however, he too has responded well to lithium.

Although schizophrenia and manic-depression can devastate patients' lives, the disorders do not preclude the performance of highly creative work. Schizophrenic patients confined in institutions have occasionally produced extraordinary works of graphic art (see Figure 8.1). Manic-depressive illness often occurs in conjunction with extraordinary talent, even genius, in politics and military leadership, as well as in literature and music and the other performing arts. Among those thought to have had the disorder are William Blake, Lord Byron, Virginia Woolf, Robert Schumann, Oliver Cromwell and Winston Churchill. Many observers have suggested that extremes of mood and changes in outlook may spur creativity; they also speculate that the energy and facility of thought that typify the milder stages of mania can be a source of creativity.

Even though schizophrenia and severe mood disorders manifest themselves as intangible mental experiences, they are biologically determined to a major degree. (Only a few of the biological discoveries can be discussed in this brief chapter.) The first evidence of determinants came early in this century, when genetic studies showed that both schizophrenia and manic-depressive illness ran in families. Most workers discounted these correlations, however, on the grounds that families share environment as well as genes. To consider the two factors in isolation, researchers turned to adoptees, who, once adopted, have environmental families that are different from their genetic families.

In the best-known study, begun in the 1960s, Seymour S. Kety and his colleagues at the National Institute of Mental Health and at a psychological institute in Scandinavia identified schizophrenics adopted in infancy and traced their biological relatives through the adoption register. The study indicated that biological relatives had an increased risk of developing the illnesses but adoptive relatives did not. The control group—biological relatives of nonpsychotic adoptees—faced no excess risk of schizophrenia or of any mental illness resembling it.

Twin studies are also revealing because different types of twins vary widely in their genetic relatedness. When schizophrenia or bipolar disease develops in one twin, the chance that it will develop in the other is much greater in identical twins, who share all their genes, than in fraternal twins, who share only about half. Moreover, although about half of the identical twins of schizophrenics never develop the illness, the children of even the well twins are at increased risk. These correlations imply two things. The risk of illness rises with increasing genetic similarity, but even a perfect identify of genes does not produce a perfect correspondence. Some environmental factor, or interaction of genes with the environment, must therefore push susceptible people over the threshold of illness. Studies have already implicated one possible factor: prenatal exposure to the influenza virus.

Mood disorders also stem from the interaction of genes with some aspect of the environment. Rates of major depression in every age group have steadily increased in several of the developed countries since the 1940s. This trend was first spotted some 10 years ago in an epidemiological study in Sweden. A similar increase in suicide over the same four decades occurred in Alberta, Canada. These findings have been firmly established as birth-cohort effects: suicide rates among 15- to 19-year-olds, for instance, were 10 times higher for those born in the late 1950s than for those born in the early 1930s. Similar birth-cohort increases appeared, over these decades, in suicide and unipolar disorder in the U.S., in bipolar disorder in the U.S. and Switzerland, and in alcoholism in males in the U.S. (see Figure 8.2).

Rates of depression, mania and suicide continue to rise as each new birth cohort ages, a pattern that harbors ominous public health consequences. Such birth-cohort effects are even more pronounced in the relatives of patients than in the general population—in other words, at comparable ages, the children of patients are far more susceptible to these disorders than are their ill parents' siblings. This relation clearly implies an interaction between genes and some environmental factor, which must have been changing continuously over the past few decades. The factor remains a mystery.

The biological abnormalities that genes and environment somehow put into motion were quite mysterious until the 1970s, when new imaging technologies allowed physicians to visualize the living brain in great detail.

One imaging technique, computerized tomography (CT scanning), was first applied to the brains of schizophrenic patients in 1978 by Eve C. Johnstone and her colleagues at the Clinical Research Centre in Middlesex, England. They observed that the lateral cerebral ventricles were much larger than in normal subjects. If the ventricles or the spaces between convolutions are enlarged, one can conclude there has been a failure of development or a loss of brain tissue. Other x-ray evidence confirmed this conclusion by showing less tissue and more fluid-filled spaces around the convolutions in the cerebral cortex.

Another technique, magnetic resonance imaging (MRI), confirmed the ventricular enlargement. Daniel R. Weinberger's group at the National Institute of Mental Health used MRI to compare identical twins in which one twin had schizophrenia and the other did not. In 12 of 15 sets of such twins, the schizophrenic one had the larger cerebral ventricles. Relative diminution of specific brain structures has been demonstrated, too, in autopsies and by MRI scans of schizophrenic patients. The most striking

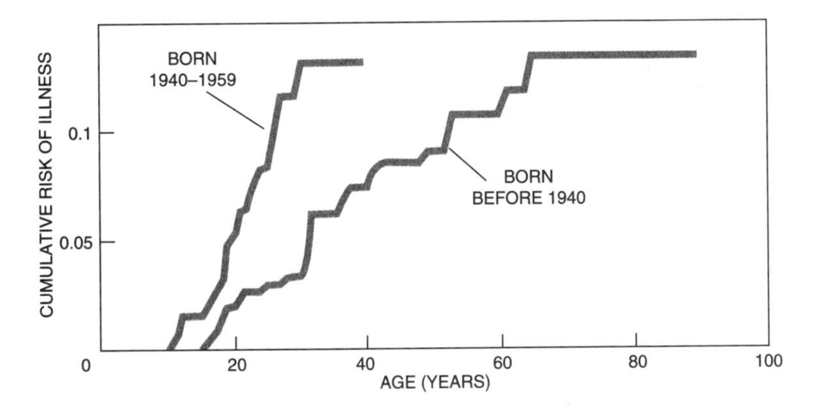

Figure 8.2 ALARMING GENERATIONAL TREND suggests that some environmental factor is increasing the incidence of mood disorders in people who are genetically susceptible to it. These graphs compare two groups of relatives of bipolar patients—one born before 1940, the other born later. At each age at which a comparison is possible, the later cohort is at far higher risk of developing bipolar illness or related psychosis.

examples of such diminution appear in the hippocampal region, part of the limbic system in the temporal lobe of the cerebrum, which modulates emotional response, memory and other functions (see Figure 8.3).

The new imaging devices also showed functional abnormalities for the first time. In 1974 David H. Ingvar of University Hospital in Lund, Sweden, found reduced blood flow in the frontal cerebrum of schizophrenic patients, implying decreased neuronal activity there. This finding has since been corroborated many times (see Figure 8.4).

Weinberger's group presents evidence linking both structural and functional abnormalities in the brain to a schizophrenic cognitive trait. They found that normal subjects show increased blood flow in the prefrontal cerebral cortex while taking the Wisconsin Card Sort, a test of working memory and

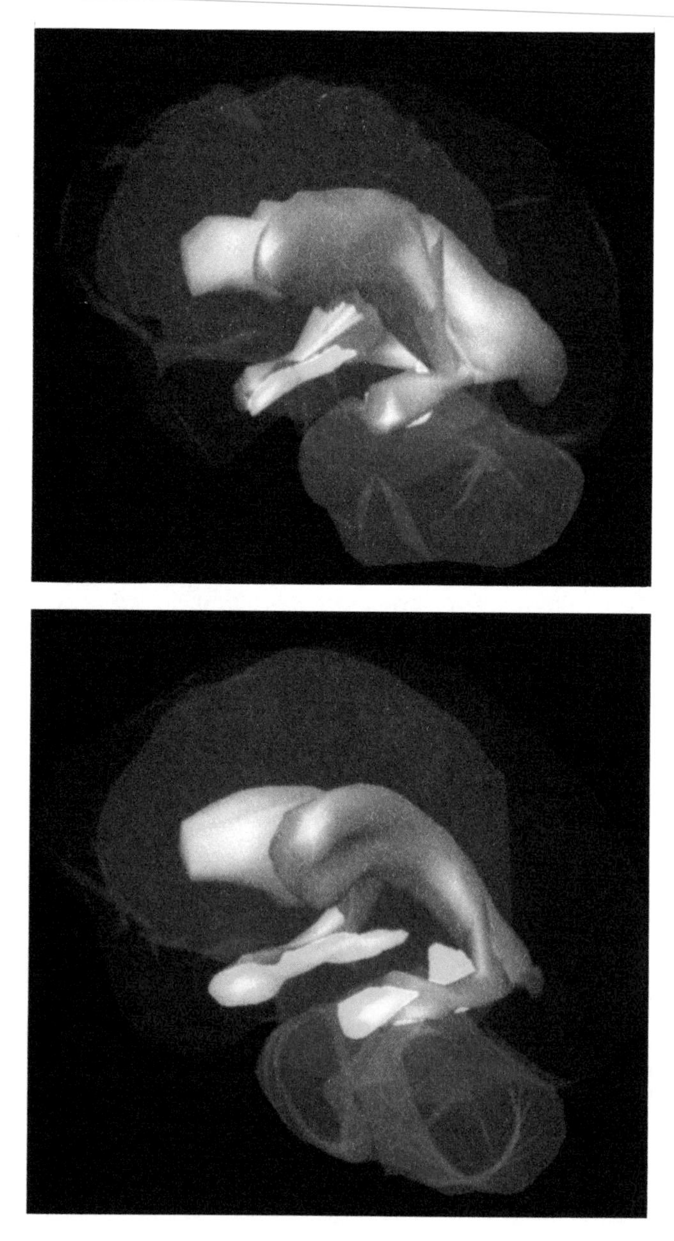

Figure 8.3 STRUCTURAL CHANGES appear in the shrunken hippocampus (*yellow*) and enlarged, fluid-filled ventricles (*gray*) of the brain of a schizophrenic patient (*top*), as contrasted with that of a normal volunteer (*bottom*). These three-dimensional MRI reconstructions were made by Nancy C. Andreasen of the University of Iowa.

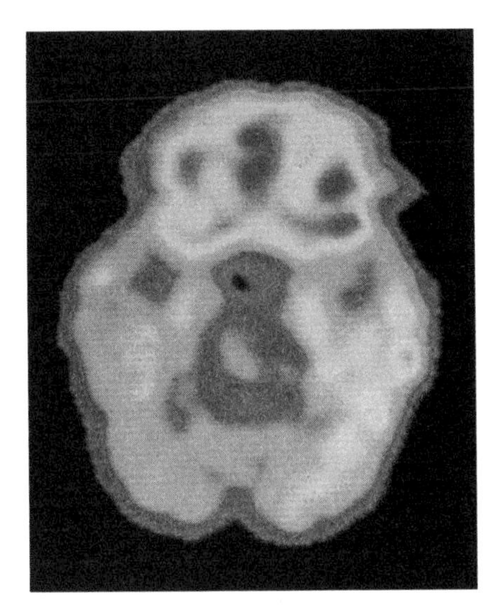

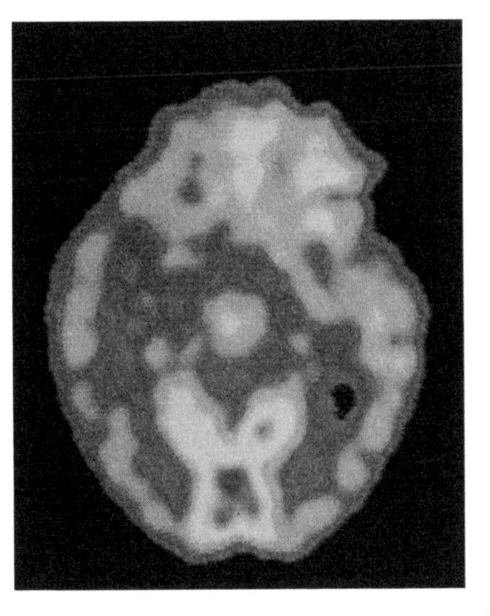

Figure 8.4 FUNCTIONAL DEFICIENCY is evident in positron emission tomographic scans made while the subjects were taking a test of vigilance. The test enhances prefrontal cortex metabolism in the normal volunteer (*top*) but not in the schizophrenic patient (*bottom*). The data were provided by Monte S. Buchsbaum of the University of California at Irvine.

abstract thinking, whereas schizophrenic subjects show less of an increase in flow and do worse on the test. Moreover, those schizophrenic patients whose hippocampal structures are the smallest show the greatest deficit in prefrontal blood flow. The hippocampus connects to the prefrontal cortex, which manages working memory in primates (see Chapter 6, "Working Memory and the Mind," by Patricia S. Goldman-Rakic).

Postmortem studies of schizophrenic patients have also uncovered abnormalities in the number of brain cells and in their organization, particularly in the temporal lobe. Yet the tissue shows none of the scarring one would expect from an infection, nor do the abnormalities progress over time. Some researchers therefore speculate that the abnormalities stem from a developmental disorder, perhaps a failure of the growth of neurons and the development of their connections, or from a disturbance in the "pruning" of neurons that normally occurs between the ages of three and 15 (see Chapter 2, "The Developing Brain," by Carla J. Shatz).

How might such abnormalities cause the symptoms of schizophrenia? While conducting surgery on nonschizophrenic patients, Wilder G. Penfield of the Montreal Neurological Institute discovered that certain structures in the brain are related to hallucinations. (Such patients are often kept awake so that they can help ascertain the functions of brain tissues near the field in which the surgeon is operating.) Penfield found that when he touched his diagnostic electrode to the temporal lobe he often elicited in his patients sights and sounds resembling hallucinations.

Further research showed that the frontal parts of the temporal cortex receive highly processed and filtered sensory information from other parts of the cortex. That information eventually reaches the limbic system and other structures that mediate emotional response, or affect. Perhaps, then, some overactivation of the temporal cortex or abnormalities in the filtering process produce the common experiences of schizophrenic patients: auditory hallucinations and a sense of being overwhelmed by all the senses.

The first effective medications for schizophrenia and depression were discovered serendipitously, without any knowledge of their effects on brain chemistry. Chlorpromazine was developed in the 1950s as a surgical anesthetic but turned out to alleviate the symptoms of both schizophrenia and mania. It thus became the first widely used antipsychotic drug. Scientists then used it as a model for the synthesis of imipramine, which they expected would also serve as an antipsychotic agent. Instead it turned out to be very effective in the treatment of depression. Lithium was introduced into the treatment of manic-depressive illness after John Cade, an Australian psychiatrist, noted in 1949 that lithium salts sedated rodents in his laboratory.

Insight into the way antidepressant agents act began with the study of reserpine, a drug derived from the plant *Rauwolfia serpentina*, used in traditional medicine in India. Reserpine was one of the first effective medications for high blood pressure. Physicians noted, however, that the drug sometimes brought on severe depression in patients—a few even committed suicide.

Biochemists discovered that reserpine depletes certain neurotransmitters classed as monoamines, among them norepinephrine, dopamine and serotonin. All the antidepressant drugs known in the mid-1960s effectively concentrated these monoamines in the synapse, either by inhibiting their metabolic breakdown or by preventing cells from reabsorbing them from the synaptic space (a process known as reuptake).

This pattern led Joseph J. Schildkraut, then at the National Institute of Mental Health, to propose in 1965 that depression was associated with a reduction in synaptic availability of catecholamines (norepinephrine and dopamine), particularly dopamine, and mania with an increase of catecholamines. Nevertheless, there are antidepressant drugs, such as iprindole, that are associated with no observable change in norepinephrine reuptake or metabolism.

Biochemical pharmacologists therefore looked beyond the neurotransmitters to the synaptic receptor molecules that bind with them. The workers knew that norepinephrine has several pharmacologically distinct receptors, called adrenoceptors, but their experiments in binding various antidepressant drugs to the receptors produced no consistent change.

Then, in 1975, Fridolin Sulser's laboratory at Vanderbilt University found an answer by looking not at binding itself but at the intracellular response that one type of binding elicits. They studied how norepinephrine stimulates beta-receptors, a class of adrenoceptors that mediates the release of cyclic adenosine monophosphate inside the nerve cell. This molecule then serves as a second chemical messenger. But after long-term administration of certain antidepressants, including iprindole, this secondary response consistently decreases. Virtually all antidepressants, including those discovered after the finding was published, produce this result. So does electroconvulsive therapy, a very effective treatment for depression in which shocks are administered to the cerebrum to induce artificial seizures.

A number of receptors for each of the monoamines exist, and new receptors continue to be discovered. With each discovery, workers learn more about how antidepressants alter these receptors and their second-messenger systems. It turns out that many, but not all, antidepressants produce dysregulations in other receptors, including postsynaptic and presynaptic adrenoceptors and certain subclasses of the dopamine and serotonin receptors.

These multiple actions of therapeutic drugs suggest that many kinds of biological defects may play a part in manic-depressive illness. Possible defects in neurotransmission include abnormalities in receptors and related molecules; various components of the second-messenger pathways; various proteins that modulate ion transport and indirectly increase or decrease activity in second-messenger systems; and various G proteins, which couple to receptors and stimulate or inhibit intracellular second messengers [see "G Proteins," by Maurine E. Linder and Alfred G. Gilman; SCIENTIFIC AMERICAN, July 1992]. So far, however, no one has found direct evidence of any such molecular abnormality in patients.

Episodes of illness gradually become more frequent and more severe in many manic-depressive patients. This pattern of deterioration suggested to Robert M. Post of the National Institute of Mental Health an analogy with an experimental process known as kindling. Scientists kindle convulsions in rodents by stimulating the animals' brains with electricity to induce seizures. Each repetition of the experiment lowers the electrical threshold for the next seizure, leading finally to spontaneous seizures.

Post proposed that manic-depressive illness progresses in a similar fashion, each episode facilitating the next one. This mechanism would account for both the progression of the illness and the deleterious effects of interrupting treatment with lithium or anticonvulsant medication. After such an interruption, patients may fail to respond to a resumption of the medication, even if it had been effective earlier.

One can also infer aspects of the biology of schizophrenia from the biochemical action of the therapeutic neuroleptic drugs, which include chlorpromazine. Arvid Carlsson of the University of Göteborg sought to explain why these drugs cause animals to produce increased quantities of breakdown products of dopamine. He suggested the effect was a compensatory response of the presynaptic neuron to a postsynaptic blockade (see boxed figure "Medicines for Mental Disorders").

As the different molecular and pharmacologic forms of dopamine receptors became known, the D_2 dopamine receptor emerged as the principal site of action of antipsychotic medications. Some of the drugs seem to work by means of interactions with other neurotransmitter systems: among these interactions are the balance between the neural pathways containing the D_1 and D_2 dopamine receptors and the balance between pathways containing certain serotonin receptors ($5HT_2$) and the D_2 dopamine receptors.

Maria and Arvid Carlsson recently proposed that schizophrenia is characterized by the disruption of a

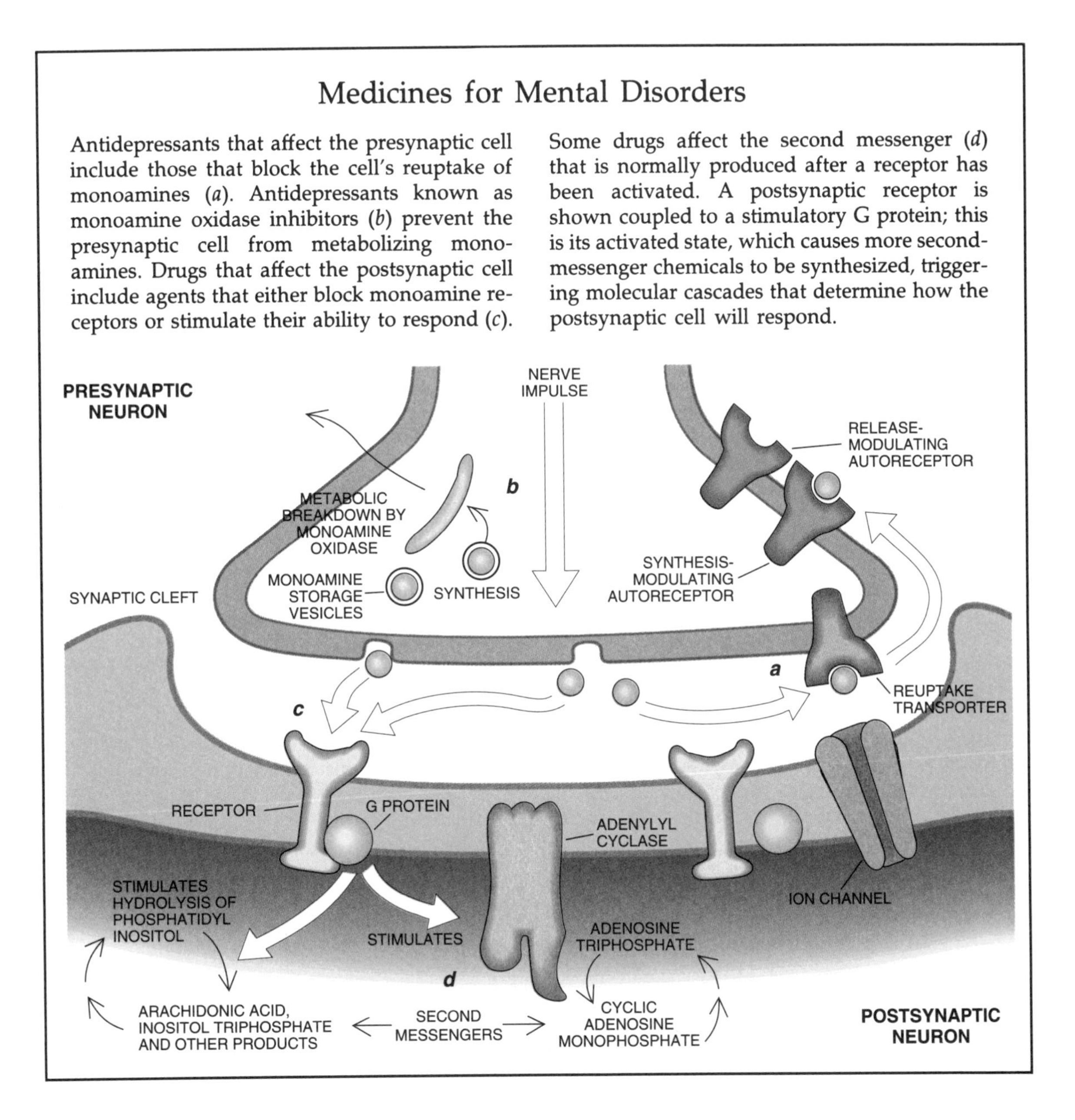

Medicines for Mental Disorders

Antidepressants that affect the presynaptic cell include those that block the cell's reuptake of monoamines (*a*). Antidepressants known as monoamine oxidase inhibitors (*b*) prevent the presynaptic cell from metabolizing monoamines. Drugs that affect the postsynaptic cell include agents that either block monoamine receptors or stimulate their ability to respond (*c*). Some drugs affect the second messenger (*d*) that is normally produced after a receptor has been activated. A postsynaptic receptor is shown coupled to a stimulatory G protein; this is its activated state, which causes more second-messenger chemicals to be synthesized, triggering molecular cascades that determine how the postsynaptic cell will respond.

balance between dopamine neurons originating in the midbrain and glutamate neurons originating in the cerebral cortex. The imbalance could be an excess of dopamine or a deficit of glutamate, or both. A reduction in glutamate neurons would be consistent with the apparent cortical atrophy seen in schizophrenia. This theory fits with the effects of known psychosis-producing drugs of abuse: PCP, a hallucinogen, blocks glutamate receptors, whereas amphetamine, which can produce psychosis with chronic use, stimulates dopamine release.

Studies of the biochemical action of these drugs and related clinical research have brought us to an era of rationally based design of medications. Once neuropharmacologists understood that chlorpromazine works by blocking dopamine receptors, they were able to synthesize haloperidol, which strongly blocks dopamine receptors but has little effect on other receptors. Similarly, after scientists found that the antidepressant imipramine blocks reuptake of the neurotransmitter serotonin, they were able to design fluoxetine, which specifically blocks serotonin reuptake but has very little effect on the reuptake of other monoamines.

Pharmacology and neurobiology continue to feed off each other. Over the past few years, clinical trials have shown that clozapine helps about 30 percent of those schizophrenic patients who do not respond to other antipsychotic medications or who develop intolerable side effects. This drug's unusual properties, such as its specific interaction with certain dopamine and serotonin receptors (D_4 and $5HT_2$), may help in the effort to understand schizophrenia itself.

Depression turns out to involve hormonal systems of a much wider scope than had been realized. Cortisol, a hormone secreted by the adrenal glands, constitutes the main circulating steroid associated with stress in humans. Many severely depressed patients show persistently elevated blood cortisol, implying a malfunction in the system that normally governs it.

George P. Chrousos of the National Institute of Child Health and Development and Philip W. Gold of the National Institute of Mental Health interpret this failure as the result of a prolonged activation of the brain's stress system (see Figure 8.5). This system—a complex of neuronal, hormonal and immunologic responses—comes into play when some stress provokes the brain, causing its hypothalamic centers to secrete corticotropin-releasing hormone (CRH). This factor in turn stimulates the pituitary gland—just under the brain—to produce the hormone adrenocorticotropin, which circulates to the adrenal glands and stimulates their release of cortisol. This process normally turns itself off when the excess cortisol reaches its receptors (glucocorticoid receptors) in the brain and suppresses CRH production there. In depressed patients, however, Gold found that production of CRH is excessive and that this suppression fails.

The CRH-producing neurons of the hypothalamus are principally regulated by neurons containing norepinephrine, which originate in the hindbrain. These CRH and norepinephrine neurons serve the stress system as central stations. Each set of neurons stimulates the other. In addition, each responds similarly to many neurotransmitters and peptide modulators of neurotransmission. Because many antidepressant drugs affect these neurotransmitters, they must also influence the regulation of the stress system.

The stress system of the brain sets the level of arousal and the emotional tone, alters the ease with which various kinds of information can be retrieved and analyzed, and aids in the initiation of specific actions. All these functions are disordered in depressed patients, and as a result they become sad, have trouble concentrating and become incapable of making decisions.

The anatomy of the stress system starts with the locus coeruleus in the hindbrain (the major source of norepinephrine-producing neurons) and the paraventricular nucleus of the hypothalamus (the brain's major CRH-producing region). From there the connections reach into the cerebrum; these connections include dopamine-producing neurons that project into the mesolimbic dopamine tract, which helps to control motivation, reward and reinforcement. A connection of CRH neurons to the amygdala and hippocampus is important for memory retrieval and emotional analysis of information pertinent to the environmental events that induced the stress.

The general concepts of stress system dysregulation apply to many psychiatric and other diseases, and it will require a considerable amount of basic and clinical research to determine whether a causal relation exists between mood disorders and this kind of stress response dysregulation.

Molecular genetics can test hypotheses on the biology of these diseases because the predisposition to them is almost certainly inherited. The task is

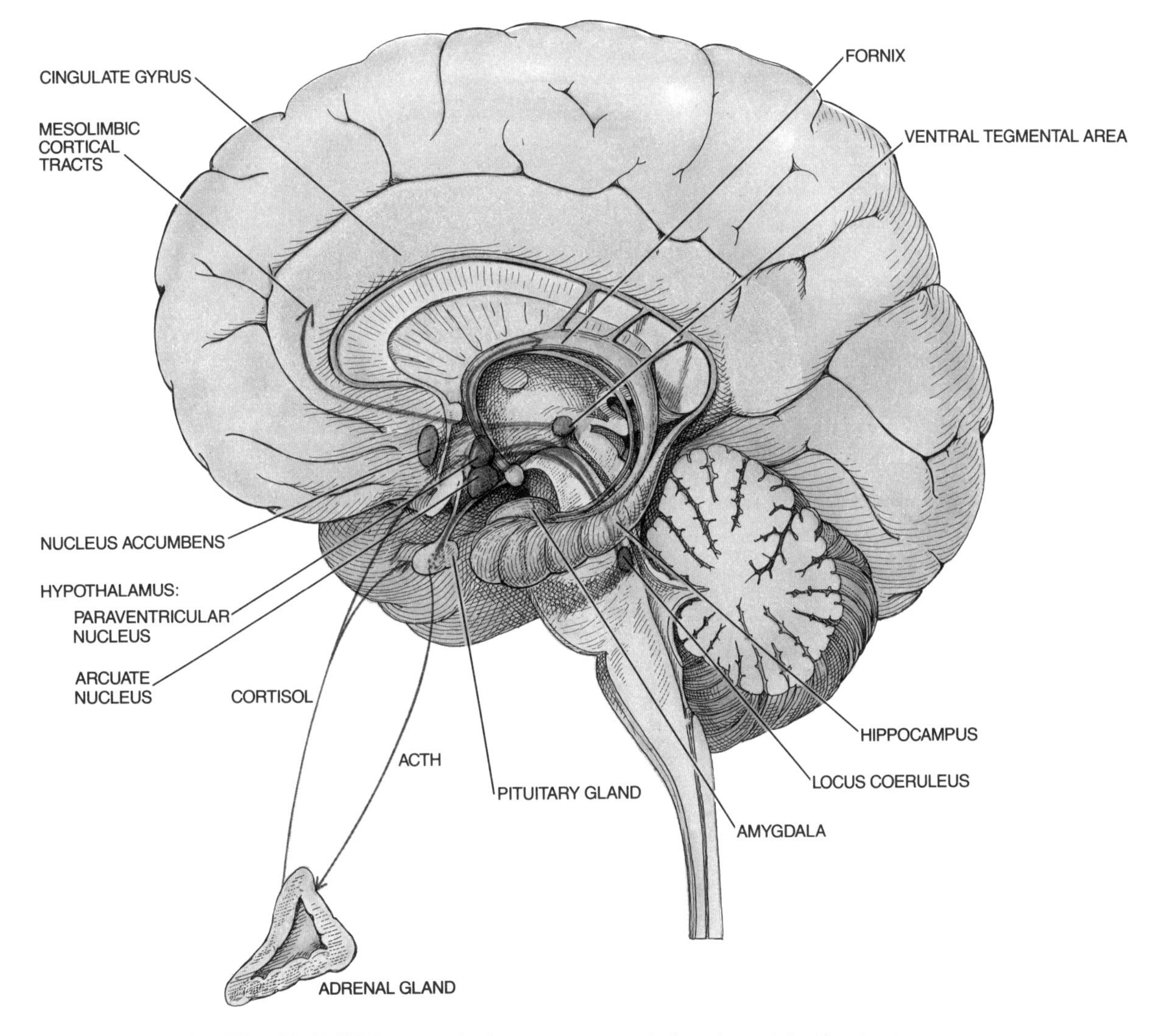

Figure 8.5 BRAIN STRESS SYSTEM extends from the hindbrain and hypothalamus to cerebral destinations inside and outside the limbic system (*blue*). It includes neurons containing norepinephrine (*dark blue*), CRH (*red*) and dopamine (*green*). When activated, the system affects mood, thought and, indirectly, the secretion of cortisol by the adrenal glands. Deactivation normally begins when cortisol binds to hypothalamic receptors. But in depression this shutdown fails, producing chronic activation.

made difficult, however, by the complexity of inheritance in schizophrenia and manic-depression. Neither illness is inherited through one dominant or recessive gene. Analysts must take into account that illness might result from coordinated actions of several genes at several different locations (loci) or, alternatively, from genetic heterogeneity (in which the same illness can be caused by a mutation at any one of several loci).

There are two major strategies for finding pathogenic genes. One can systematically search each chromosome, or one can investigate a candidate gene—such as that for a given receptor—which is known to code for proteins related to the disease.

DNA markers now exist for nearly every segment of every chromosome. In any family, each parent contributes to the child a single segment from either one or the other chromosome in a given pair of chromosomes. The illness is linked to a marker location on the genetic map if, and only if, one ancestral chromosome segment is consistently inherited with illness throughout a pedigree. Whenever there is a linkage to the marker, one can be sure that a gene for the illness resides somewhere on that chromosome segment.

Such mapping has consistently shown that a proportion of the families of Alzheimer's patients have illness linked to markers on the long arm of chromosome 21. No such linkages have been demonstrated conclusively for manic-depressive illness or schizophrenia. One widely publicized study of a large Amish pedigree linked manic-depressive illness to markers on chromosome 11; another study of a series of pedigrees in Iceland and England linked schizophrenia to markers on chromosome 5. But later analyses led the investigators to withdraw their conclusions, and no other researchers have confirmed the findings.

Linkage to manic-depressive illness has been reported more than once at the tip of the long arm of the X chromosome, but the linkage remains controversial. One can expect more definitive results from several large-scale international efforts, now under way, to scan the entire gene map of families touched by either schizophrenia or manic-depressive illness.

Many genes encoding the molecules involved in neurotransmission are candidates for the defects underlying manic-depressive illness or schizophrenia. A study of several such candidates for manic-depressive illness was performed on a series of 20 pedigrees at the National Institute of Mental Health by Margret R. Hoehe, Sevilla D. Detera-Wadleigh, Wade H. Berrettini, Pablo V. Gejman and one of us (Gershon). The group tested structural genes for many of the receptors we have described, including norepinephrine (three alpha-receptor genes and two beta-receptor genes), dopamine (D_2 and D_4) and corticosteroid receptors, and the gene of a G protein subunit ($G_{s\alpha}$). Other investigators have studied the D_2 dopamine receptor in schizophrenia pedigrees. Linkage was strongly excluded for each of these genes.

When linkage of a candidate gene to illness can be ruled out, one can conclude that no mutation in the candidate gene determines inheritance of the susceptibility to the illness. The only qualification to this general rule is the statistical and technical limits on our power to detect or rule out linkage. Methods other than linkage can also scan for mutations in candidate genes, and many genes remain to be examined.

We expect our understanding of the biology of schizophrenia and mood disorders to expand dramatically, fueled by the impressive advances in neurobiology, cognitive neuroscience and genetics. Precise diagnostic tests for persons at risk for illness, treatments based on knowledge of molecular alterations that lead to illness, understanding of how environmental events interact with the brain to produce illness and, eventually, the development of gene therapy are all goals that may be achieved.

Aging Brain, Aging Mind

Late in life the human brain suffers attrition of certain neurons and undergoes chemical alterations. Yet for many people, these changes do not add up to a noticeable decline in intelligence.

• • •

Dennis J. Selkoe

Contemplate senescence, as Shakespeare did. In *As You Like It*, his memorable character Lord Jaques enumerates seven ages of man, concluding with this sad description:

Last scene of all, that ends this strange
eventful history,
Is second childishness and mere oblivion.

For many of us, as for the melancholy Jaques, the prospect of aging continues to conjure images of inexorable, devastating decline, a slow march toward mindlessness and mortality. But is severe deterioration of the brain—and thus the mind—inevitable?

The answer is no (see Figure 9.1). Admittedly, research indicates that as youth fades, certain molecules and cells in the brain become increasingly impaired or disappear. Some of the changes can undoubtedly disrupt cognition if they accumulate past critical thresholds. Yet studies of human behavior suggest that a mind-eroding buildup of damage is by no means an automatic accompaniment of longevity.

Older adults who truly lose their minds probably do so because a specific disease markedly accelerates or is superimposed on the aging process. In developed nations, the leading cause of senile dementia—loss of memory and reason in the elderly—is Alzheimer's disease. Other causes include the occurrence of multiple strokes or Parkinson's disease.

Physicians cannot always distinguish between older people who suffer from minor, relatively stable forgetfulness and those who are in the early stages of Alzheimer's disease or another progressive, dementing disorder. Ongoing research into normal aging and into disorders of the mind will both enable doctors to make those distinctions and give rise to palliative and preventive therapies. For most students of the aging brain, the ultimate goal is to enhance the quality of its function in old age, not necessarily to prolong life, although the latter could certainly result from the former.

Scientists who study the structural and chemical changes that typify the aging brain in the absence of disease find that the changes are heterogeneous, like the brain itself. The brain consists not only of diverse neurons (the signal-conveying cells) but also of varied glial cells (which help to support and repair neurons) and of blood vessels. Certain subsets of cells and areas of the brain are more prone to age-related damage than others. And the time of onset and the mix and extent of physical alterations,

as well as the effects on intellect, can differ dramatically from person to person. In general, however, it seems safe to say that most of the structural and chemical modifications I will discuss become apparent in late middle life, that is, in the fifties and sixties. Some of them become pronounced after age 70. Because there probably is no unifying mechanism that underlies all senescence (age-related dysfunction of cells and molecules) in the brain, it seems unlikely that investigators will find a singular elixir that will retard or reverse every decline.

Age-associated changes have been most studied in neurons, which in general do not multiply after birth (see Figure 9.2). As individuals grow older, their overall number of brain neurons decreases, but the pattern is by no means uniform. For example, very few neurons disappear from areas of the hypothalamus that regulate the secretion of certain hormones by the pituitary gland.

In contrast, many more nerve cells tend to vanish from the substantia nigra and locus coeruleus, which are specialized populations of cells in the brain stem. Parkinson's disease can decimate some 70 percent or more of the neurons in those areas, seriously disrupting motor function. Aging alone usually eliminates many fewer cells, although older individuals who exhibit mild symptoms reminiscent of Parkinson's disease—decreased flexibility, slowness of movement and a stooped, shuffling gait—may have lost up to 30 or 40 percent of the original complement.

Parts of the limbic system, including the hippocampus, undergo variable amounts of cell death as well. (The limbic system is central to learning, memory and emotion.) Researchers have estimated that about 5 percent of neurons in the hippocampus disappear with each decade in the second half of life. This calculation suggests that about 20 percent of neurons are lost in that period. The attrition is patchy, though; some hippocampal areas show no significant decline (see Figure 9.3).

Even when neurons themselves survive, their cell bodies and their complex extensions known as axons and dendrites (or, collectively, as neurites) may atrophy. Neurons bear a single axon that relays signals to other neurons, often a distance away. Dendrites, which are more abundant and are found in large branching arbors, generally receive signals from other neurons.

Figure 9.1 GEORGE BERNARD SHAW, who died in 1950 when he was 94, wrote several plays in his nineties. A major goal of research into the aging brain is to increase the number of people who retain mental vigor throughout life.

Cell-body and neuritic atrophy commonly occur with aging in a number of brain areas important to learning, memory, planning and other complex intellectual functions. Large neurons, in particular, shrink in parts of the hippocampus and the cerebral cortex. And cell bodies and axons can degenerate in certain acetylcholine-secreting neurons that project from the basal forebrain to the hippocampus and diverse areas of the cortex. Acetylcholine is one of various neurotransmitters by which neurons convey signals to one another.

Yet not all neuronal changes are necessarily destructive. Some may represent attempts by surviving neurons to compensate for loss or shrinkage of other neurons and their projections. Indeed, Paul D. Coleman, Dorothy G. Flood and Stephen J. Buell of the University of Rochester Medical Center have observed a net growth of dendrites in some regions of the hippocampus and cortex between middle age (the forties and fifties) and early old age (the early seventies), followed by a regression of dendrites in late old age (the eighties and nineties). These investigators postulate that the initial dendritic growth reflects an effort by viable neurons to cope with the age-associated loss of their neighbors. Apparently, this ability to compensate fails in very old neurons. Studies of adult rats show a similar capacity for growth; longer and more complex dendrites appear in the visual cortex after the animals are exposed to visually stimulating environments.

Such findings are encouraging. They suggest both that the brain is capable of dynamic remodeling of its neuronal connections, even in the later years, and that therapy of some kind might augment this plasticity. On the other hand, the functionality of the dendrites that appear in old age has yet to be determined.

In addition to changes in their number and in the structure of their cell bodies and neurites, neurons can undergo alteration of their internal architecture. For example, the cytoplasm of certain cells of the hippocampus and other brain areas vital to memory and learning can begin to fill with bundles of helically wound protein filaments known as neurofibrillary tangles (see Figure 9.4). An abundance of such tangles in these and other brain areas is believed to contribute to the dementia of Alzheimer's

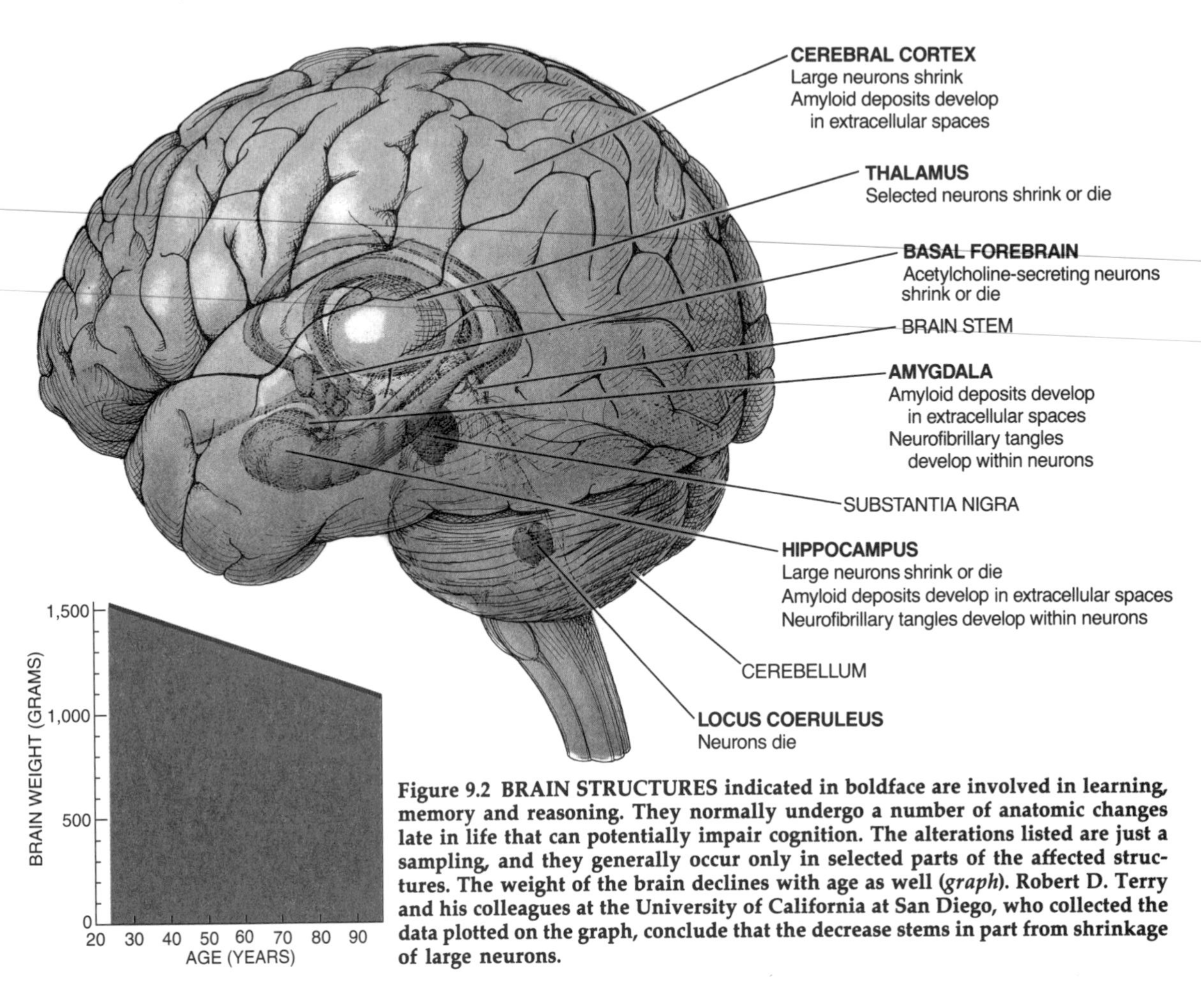

Figure 9.2 BRAIN STRUCTURES indicated in boldface are involved in learning, memory and reasoning. They normally undergo a number of anatomic changes late in life that can potentially impair cognition. The alterations listed are just a sampling, and they generally occur only in selected parts of the affected structures. The weight of the brain declines with age as well (*graph*). Robert D. Terry and his colleagues at the University of California at San Diego, who collected the data plotted on the graph, conclude that the decrease stems in part from shrinkage of large neurons.

disease, but the significance of small amounts in the undiseased brain is less clear. The development of tangles during aging seems to indicate that certain proteins, particularly those of the cytoskeleton, or the internal scaffolding of the cell, are being chemically modified in ways that could contribute to inefficient signaling by these neurons.

In another internal alteration, neuronal cytoplasm in many parts of the brain becomes increasingly dotted with innumerable granules containing lipofuscin, a fluorescent pigment. This pigment is thought to derive from lipid-rich internal membranes that have been incompletely digested. Again, investigators disagree over whether lipofuscin granules harm cells or are mere markers of longevity.

As is true of neurons, glial cells, which have a supporting role in brain function, also become altered. Robert D. Terry of the University of California at San Diego and other investigators have established that the type known as fibrous astrocytes increases steadily in size and number after age 60. Proliferation of these cells, which are capable of releasing diverse factors that promote neuronal and neuritic growth, is of unknown consequence. Perhaps it represents another attempt by the brain to compensate for gradual decrements in the number and structure of neurons.

Meanwhile the areas between neurons are also undergoing change. In humans, monkeys, dogs and certain other animals, the extracellular spaces of the hippocampus, cerebral cortex and other brain regions commonly accumulate moderate numbers of spherical deposits called senile plaques. These plaques, which develop very slowly, are primarily aggregates of a small molecule known as the beta-

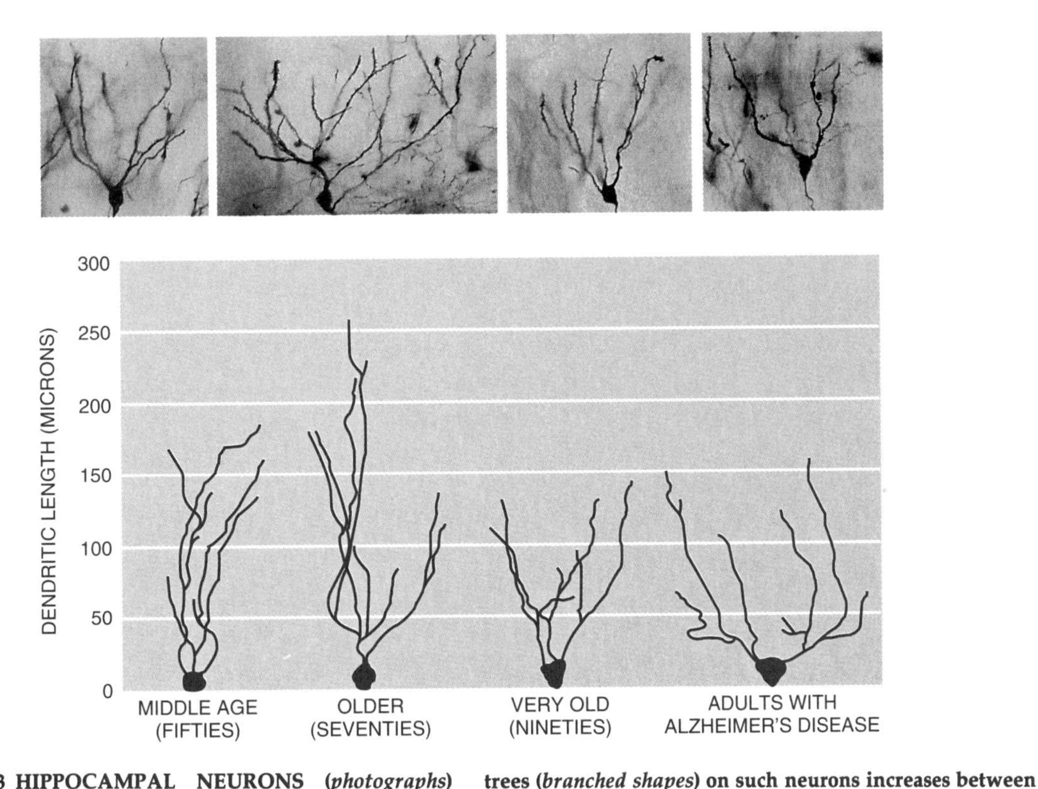

Figure 9.3 HIPPOCAMPAL NEURONS (*photographs*) have been stained in brain specimens from healthy humans in their fifties, seventies and nineties and from a patient with Alzheimer's disease (*left to right*). Dorothy G. Flood and Paul D. Coleman of the University of Rochester Medical Center find that the average length of dendritic trees (*branched shapes*) on such neurons increases between middle age and old age in healthy adults, regressing only in late old age (*graph*). The normal growth may reflect an attempt by the brain to compensate for destructive age-linked changes. Dendrites in Alzheimer's patients do not exhibit age-related growth.

amyloid protein. Amyloid protein also accumulates in scattered blood vessels in these regions and in the meninges, the connective tissue that envelops the brain.

Researchers have not fully determined which cells give rise to these protein deposits and what effect the accumulations have on nearby neurons in healthy elders. The answers should emerge soon, however, because the dramatically increased deposition of amyloid protein in patients with Alz-

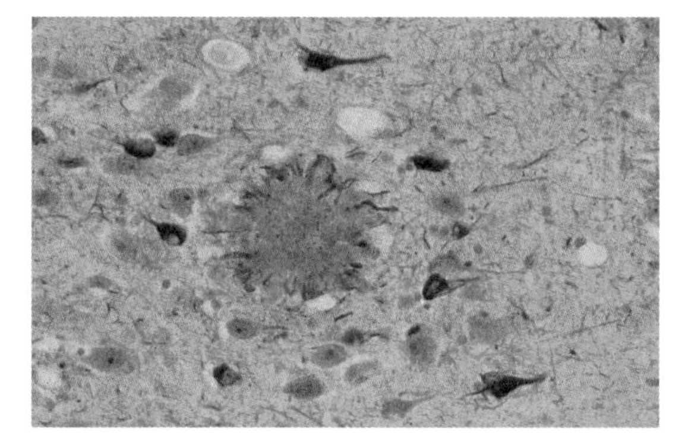

Figure 9.4 TISSUE FROM THE BRAIN of a 69-year-old man is riddled with the classic lesions of Alzheimer's disease—senile plaques and neurofibrillary tangles. The plaque visible in this fragment (*large dark-gold sphere*) consists of beta-amyloid protein and, at the periphery, damaged axons and dendrites (*dark squiggles*). The tangles, which are twisted fibers that fill the cytoplasm, make a number of cells appear blackened (*small dark blobs*). Plaques and tangles appear in the brains of healthy aged adults, but usually to a much lesser extent and in restricted regions.

heimer's disease has pushed such issues to the forefront of research.

The diverse structural alterations that occur in the aging brain result from deleterious changes in the activity or abundance of molecules that are important to the integrity and functioning of cells. One of the most venerable theories of aging holds that cells throughout the body senesce because defects slowly accrue in their DNA, the material from which genes are constructed. Genes carry the chemical instructions that inform cells precisely how to synthesize proteins. At some point, the scenario goes, damage to DNA lowers the quality or quantity of critical proteins (such as certain enzymes) in cells. Or the damage may increase the activity or amount of undesirable proteins (such as those that promote the development of cancers).

Until recently, genetic research focused almost entirely on the chromosomal DNA in the nucleus, the long strands of helical DNA that collectively store the genes for virtually all the proteins made in cells. Such work indicated that the enzymatic machinery designed to excise and repair faulty nuclear DNA becomes less efficient late in life and perhaps in the presence of certain brain diseases. Scientists have also found evidence that cellular controls governing genetic activity may be related during aging. One mechanism may involve subtle elimination of methyl groups (CH_3) from certain parts of DNA molecules [see "A Different Kind of Inheritance," by Robin Holliday; SCIENTIFIC AMERICAN, June 1989]; Offprint 1606].

In recent years, investigators have begun to suspect that DNA in a special cellular location—the mitochondria—may also contribute to senescence of the brain. Mitochondria are intracellular "power plants" that provide cells with critically needed energy. They contain their own snippet of DNA, which bears instructions for the manufacture of 13 proteins needed for energy generation. If mitochondrial DNA slowly became partially defective, the defects could result in production of damaged mitochondrial proteins or in the elimination of such proteins.

The evidence favoring a causal role for mitochondrial DNA in some of the changes associated with aging includes the finding that such DNA seems to be more susceptible to damage than is nuclear DNA. One reason may be that the DNA-repair machinery in the organelles is less effective than that in the nucleus. Mitochondrial DNA is also probably subject to more attack by highly reactive oxygenated compounds known as free radicals. Such compounds are a continual by-product of the reactions through which mitochondria produce energy. Free radicals are made during other cellular processes as well and in response to ionizing radiation. They oxidize, or add oxygen atoms to, molecules, thereby altering them.

Furthermore, investigators have discovered that a crucial enzyme encoded by mitochondrial DNA—cytochrome oxidase—declines with age in rat brains. Several researchers have also identified specific deletions in stretches of mitochondrial DNA in aged brains and in patients with certain age-linked brain disorders, such as Parkinson's disease.

Even if most nuclear and mitochrondrial genes remained unaltered and gave rise to the proper amounts of normal proteins, subsequent modifications of the proteins could lead to molecular mischief in late life. Proteins can undergo a number of distinct chemical modifications, including oxidation of certain of their component amino acids, glycosylation (addition of carbohydrate side chains) or cross-linking (formation of strong chemical bridges between proteins). Such modifications occur normally and enable proteins to carry out their functions. On the other hand, there is abundant evidence that as people age, many proteins accumulate unhelpful modifications. For example, the levels of oxidized proteins in skin cells of humans and brain cells of rats increase progressively with age. In very old rats, such proteins can account for 30 to 50 percent of the total protein content of the cell. Cells from young adults with progeria, a rare and remarkable syndrome marked by premature aging of many body tissues, contain levels of oxidized proteins that approach those found in healthy 80-year-olds.

Because enzymes are the proteins that catalyze many of the most important chemical reactions in cells, they have received the greatest scrutiny. Several enzymes that synthesize neurotransmitters or their receptors become less active as an individual ages. For some of these enzymes, postsynthetic modification may be partly at fault.

It is a cruel irony that proteases, the very enzymes responsible for degrading oxidized and other proteins, themselves undergo oxidation and loss of activity. The problem of damaged proteases, potentially bad enough on its own, may be exaggerated by a parallel decline in the enzymes superoxide dismutase and catalase. These proteins normally inactivate free radicals and defend against oxidative damage to diverse kinds of molecules. At least in rats, they become increasingly scarce in old age.

John M. Carney of the University of Kentucky and Robert A. Floyd of the Oklahoma Medical Research Foundation and their co-workers have recently provided some of the first evidence that such oxidation may lead to a loss of mental function. Comparing aged and young gerbils, they showed that significantly more proteins in the older animals were oxidized. This increase was accompanied by a decrease in activity of certain critical enzymes. Moreover, the older animals had more trouble than did the younger ones in navigating a radial maze (see Figure 9.5).

When the scientists injected the aged gerbils with a compound (*N-tert*-butyl-α-phenylnitrone) known to inactivate oxygen free radicals and thus decrease oxidation, the levels of oxidized proteins declined, and the activity of the enzymes increased to the levels characteristic of young animals. What is more, the biochemical improvements were accompanied by the restoration of maze-running skill to youthful levels. When the therapy was stopped, the amount of oxidized protein and enzymatic activity reverted to that typical of old animals.

Many important nonprotein molecules in the brain also change significantly in structure or amount during aging. There is evidence that the long chains of carbon atoms making up the lipids in membranes that envelop cells and internal organelles undergo chemical modifications. Among these is destructive oxidation by free radicals. As a result, the precise compositions of the membranes can shift, subtly altering their behavior.

For example, investigators have documented an age-related decrease in the fluidity of the membranes making up synaptosomes, tiny neuronal vesicles involved in the storage and release of neurotransmitters. And age-related changes occur in the lipid composition of the myelin that sheathes and insulates axons. Alteration of myelin can have a measurable effect on the speed and efficiency with which nerve fibers propagate electrical impulses over long distances.

The molecular changes I have discussed are but a small sampling of those that have been found in aged brains of humans and other mammals. In trying to make sense of such alterations, scientists immediately confront the problem of determining whether a phenomenon they have documented is the proverbial "cart" or "horse." For instance, there is little doubt that DNA accrues damage over the years. But does the damage result, for example, in increased oxidation of enzymes, or does oxidation occur first and lead to accumulation of DNA deficits? The likelihood is that both sequences can happen. Once many processes get started, they undoubtedly exacerbate others, triggering a complex cascade of events.

Equally important is the question of what bearing

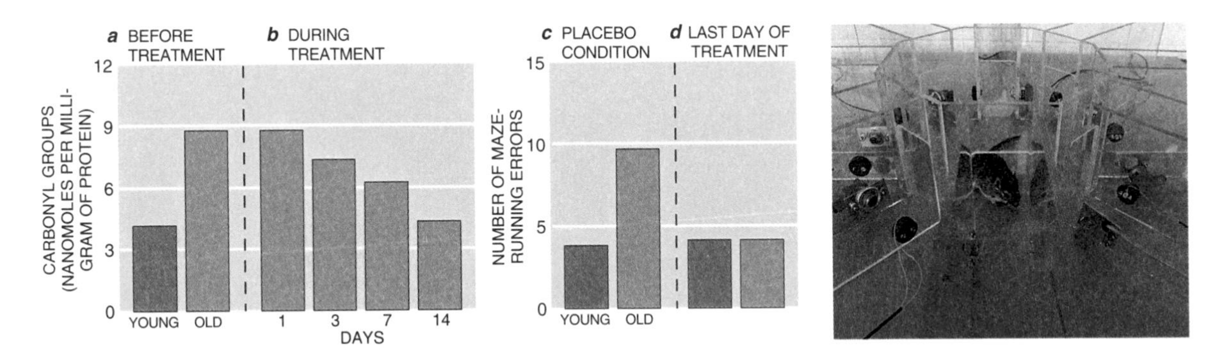

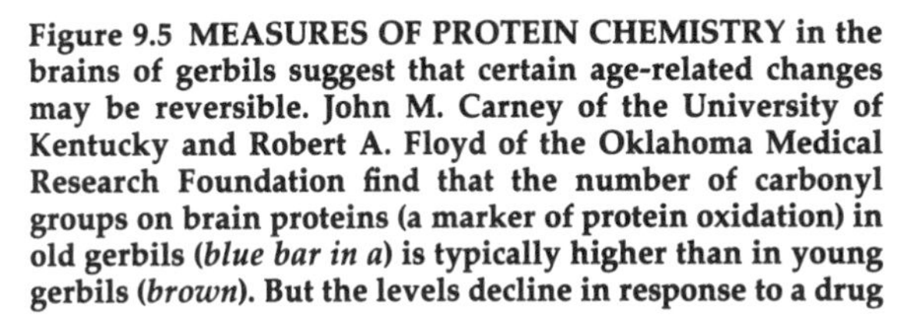

Figure 9.5 MEASURES OF PROTEIN CHEMISTRY in the brains of gerbils suggest that certain age-related changes may be reversible. John M. Carney of the University of Kentucky and Robert A. Floyd of the Oklahoma Medical Research Foundation find that the number of carbonyl groups on brain proteins (a marker of protein oxidation) in old gerbils (*blue bar in a*) is typically higher than in young gerbils (*brown*). But the levels decline in response to a drug that inactivates certain oxidizing chemicals (*b*). Geriatric gerbils are also less efficient than youngsters at negotiating a radial maze (*c*), such as the one shown in the photograph. After treatment, the number of errors drops (*d*). The apparent improvement in short-term memory encourages hope that antioxidants may one day help to somewhat protect the minds of aging humans.

all these diverse anatomic and physiological age-linked changes have on the mind. In many people, the answer may be "very little." Until scientists can record the mental functioning of large numbers of healthy people close to the time they die and then correlate such data with structural and chemical changes in their brains, any links between specific physical alterations and disruptions of the intellect will remain murky.

We do know that in people who are free of Alzheimer's disease and other specific brain-threatening disorders, the extent of anatomic and physiological alterations tends to be modest. In many studies reporting an age-related neurochemical deficit—such as reduction in the activity of a particular enzyme or in the levels of selected proteins or RNA molecules—the measures reported for elderly adults have ranged from 5 to 30 percent below those in young adults. The degree of neuronal loss in various regions of the brain falls into roughly the same range.

Although a 30 percent loss might seem quite high, such gradual declines often appear to have little practical effect on thinking. Indeed, positron emission tomographic (PET) imaging indicates that the brains of healthy people in their eighties are almost as active as those of people in their twenties (see Figure 9.6). As is true of other organs, the brain appears to have considerable physiological reserve and to tolerate small losses of neuronal function.

Epidemiological and psychological studies paint a similar picture. Estimates of the prevalence of dementia vary, but the most dire of them—derived from a door-to-door study conducted by Dennis A. Evans of Harvard Medical School and his colleagues—indicates that as a group, almost 90 percent of all people older than 65 years are free of dementia. Evans and his co-workers reported in 1991 that fewer than 5 percent of subjects aged 65 to 75 exhibited symptoms of dementia, a figure that rose to about 20 percent in the subjects aged 75 to 84. Then the number jumped to about 50 percent in people older than 85 years (which is twice as high as certain other estimates). As disturbing as the data are for people older than 75 years, the figures still indicate that a good many people escape major disturbances in cognition in the later years (see boxed figure "Dementia in the U.S.").

Analyses of performance in healthy old adults lead to a similar conclusion. For instance, Arthur L. Benton, Daniel Tranel and Antonio R. Damasio of the University of Iowa College of Medicine found that when people in their seventies and eighties remain in good health, they show only a subtle decline in performance on tests of memory, perception and language.

One decrement on which many studies agree is a reduction in the speed of some aspects of cognitive processing. Hence, septuagenarians may be unable to retrieve certain details of a particular past event quickly—say, the precise date or place—but they are often able to recall the information minutes or hours later. Given enough time and an environment that keeps anxiety at bay, most healthy elderly people score about as well as young or middle-aged adults on tests of mental performance. The more complex a task is (for example, a multistep mathematical problem), the more likely it is that an otherwise healthy elder will perform less well than does a young adult. A message of guarded optimism emerges from many investigations of normal aging: one may not learn or remember quite as rapidly during healthy late life, but one may learn and remember nearly as well.

Taken together, then, the physical, epidemiological and psychological findings suggest that mild to

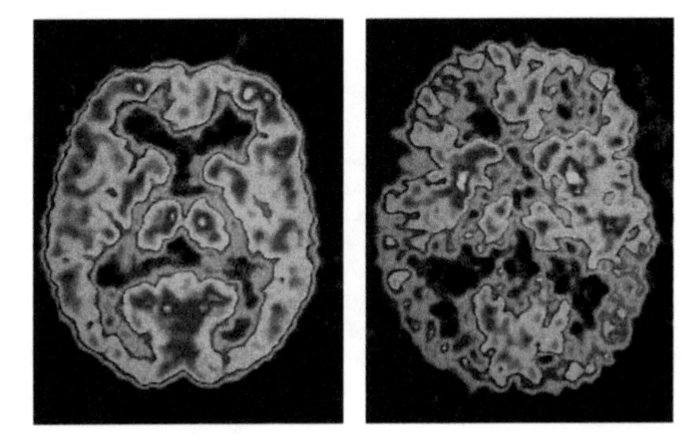

Figure 9.6 PET SCANS OF BRAINS from a healthy aged adult (*left*) and a patient with Alzheimer's disease (*right*) differ markedly. The excess of dark shading in the second image shows that brain activity is impaired.

moderate decreases in memory or speed of intellectual processing may be related to a gradual accumulation of standard anatomic and physiological brain changes that accompany aging. In contrast, dementia apparently arises from more specific and excessive changes in subsets of neurons and in neural circuits. In other words, diseases having distinct causes and mechanisms underlie senile dementia. Of course, one might well wonder why people become more prone to various debilitating brain disorders, including Alzheimer's disease, as they age. In many instances, the answers are uncertain.

Because Alzheimer's disease is by far the most common cause of severe intellectual decline in the elderly, let me briefly review the latest research into why it develops, why it becomes increasingly prevalent late in life and how it might eventually be treated. Fortunately, progress in this area has been truly remarkable of late [see "Amyloid Protein and Alzheimer's Disease," by Dennis J. Selkoe; SCIENTIFIC AMERICAN, November 1991].

Until very recently, the question of what causes Alzheimer's disease had to be answered, "We don't know." But rapid advances from many laboratories analyzing the chemistry and molecular biology of the beta-amyloid deposits has now led to identification of the first specific molecular cause of this complex and devastating disorder. In 1991 studies by Alison M. Goate and John A. Hardy and their colleagues at St. Mary's Hospital Medical School in London, and subsequently other research teams, established that particular genetic mutations are at fault in at least some instances.

These DNA mutations occur within the gene that gives rise to the beta-amyloid precursor protein, or beta-APP. This precursor includes within it the beta-amyloid protein that constitutes both senile plaques and vascular amyloid deposits. The normal functions of beta-APP have not yet been revealed, but those of us who study Alzheimer's disease have found that the precursor is made by most cells of the body. We know, too, that the mutated version somehow leads to accelerated extracellular and vascular deposition of the beta-amyloid segment. Some mutations may lead to more or faster amyloid accumulation than others. Hastened depositions, in turn, could partly explain why some people show symptoms earlier than others.

Research on Down's syndrome has contributed importantly to the new understanding. Individuals with that syndrome are born with three copies of chromosome 21 (where the beta-APP gene is located) rather than the normal two copies. They also invariably develop myriad senile plaques and neurofibrillary tangles in the fourth and fifth decades of life. Neuropathological examination of Down's patients who have died early in life has revealed that a few amorphous deposits of amyloid protein can begin to appear in the teens, decades before full-blown senile plaques and neurofibrillary tangles and clinical signs of dementia develop. That critical finding, together with the discovery of beta-APP mutations in inherited Alzheimer's disease, now makes it clear that amyloid protein deposition can serve as a seminal event in some if not all cases of Alzheimer's disease.

No one is sure how the initially inert protein leads over a long time to the extensive structural and biochemical changes in axons, dendrites, neuronal cell bodies and glial cells that ravish the minds of Alzheimer's victims. One possibility is that the protein itself remains unreactive, but as it collects over many years, it attracts other types of molecules to the deposits. These other molecules may then damage surrounding neurons and glia. Another hypothesis suggests that after the amyloid protein reaches critical concentrations, it directly damages surrounding neurons and glia or makes them more vulnerable to subtle injurious processes that can occur in the brain.

In any case, the work of several neuroanatomists —including Damasio, Bradely T. Hyman and Gary W. Van Hoesen and their co-workers at the University of Iowa and John Morrison of the Mount Sinai School of Medicine and his colleagues—has shown that a buildup of amyloid protein, combined with the formation of neurofibrillary tangles and other structural changes in neurons and their extensions, contributes to a progressive disconnection of neuronal circuits serving memory and thinking. Over years the limbic system and the association cortices, which are vital to organizing mental processes, appear to become increasingly out of touch with other neuronal areas. This disconnection helps lead to the impaired memory, judgment, abstraction and language that is all too familiar in Alzheimer's patients. Because most motor and sensory functions are spared until rather late in the disease, the changes give rise to the classical, tragic picture of a person who can walk, talk and eat but cannot make sense of the world.

In spite of recent progress, many central problems need to be addressed. How do mutations in the

Dementia in the U.S.

In 1992 the Framingham Study, which repeatedly assesses the health of a large group of subjects as they age, estimated the prevalence of dementia (*a*), including Alzheimer's disease (*b*). The Alzheimer's figures differ from those of a survey conducted in East Boston (by a team led by Denis A. Evans of Harvard Medical School), probably because the Framingham group applied a narrower definition of the disease. The Framingham Study, consistent with others, also found Alzheimer's disease to be the leading cause of persistent dementia in late life (*c*). Some currently treatable causes are listed (*d*).

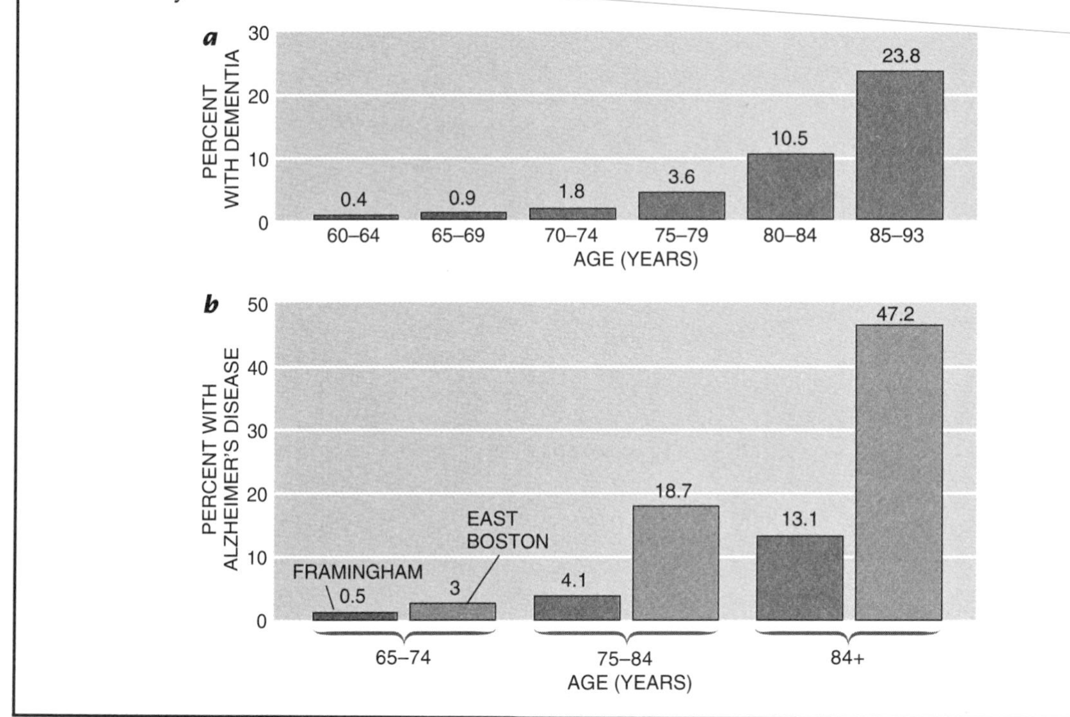

beta-APP gene lead to accelerated amyloid deposition compared with the slow rate encountered in normal aged humans? Why is such deposition largely confined to the brain, when virtually all tissues synthesize the amyloid precursor? Which cells actually secrete the devastating amyloid fragments? Why do some neurons in certain brain regions, such as the hippocampus, show striking reaction to the presence of amyloid protein, yet other areas, such as the cerebellum, exhibit little or none? Most important, how can the terrible destruction be blocked?

As these questions are being pursued, so is the problem of how to bolster and protect the aging mind. The fact that no single compound is likely to block all the potential ravages of great longevity is underscored by results of many clinical trials of vitamins, minerals and various other compounds thought to "enhance" biochemical reactions in the brain or increase blood flow. These substances have yielded little or no cognitive improvement in either demented or functional elderly people.

One reasonable "home remedy" would be to stay physically fit. Robert E. Dustman and his colleagues at the University of Utah and other investigators have demonstrated that older subjects who regularly do aerobic exercise perform better on cognitive

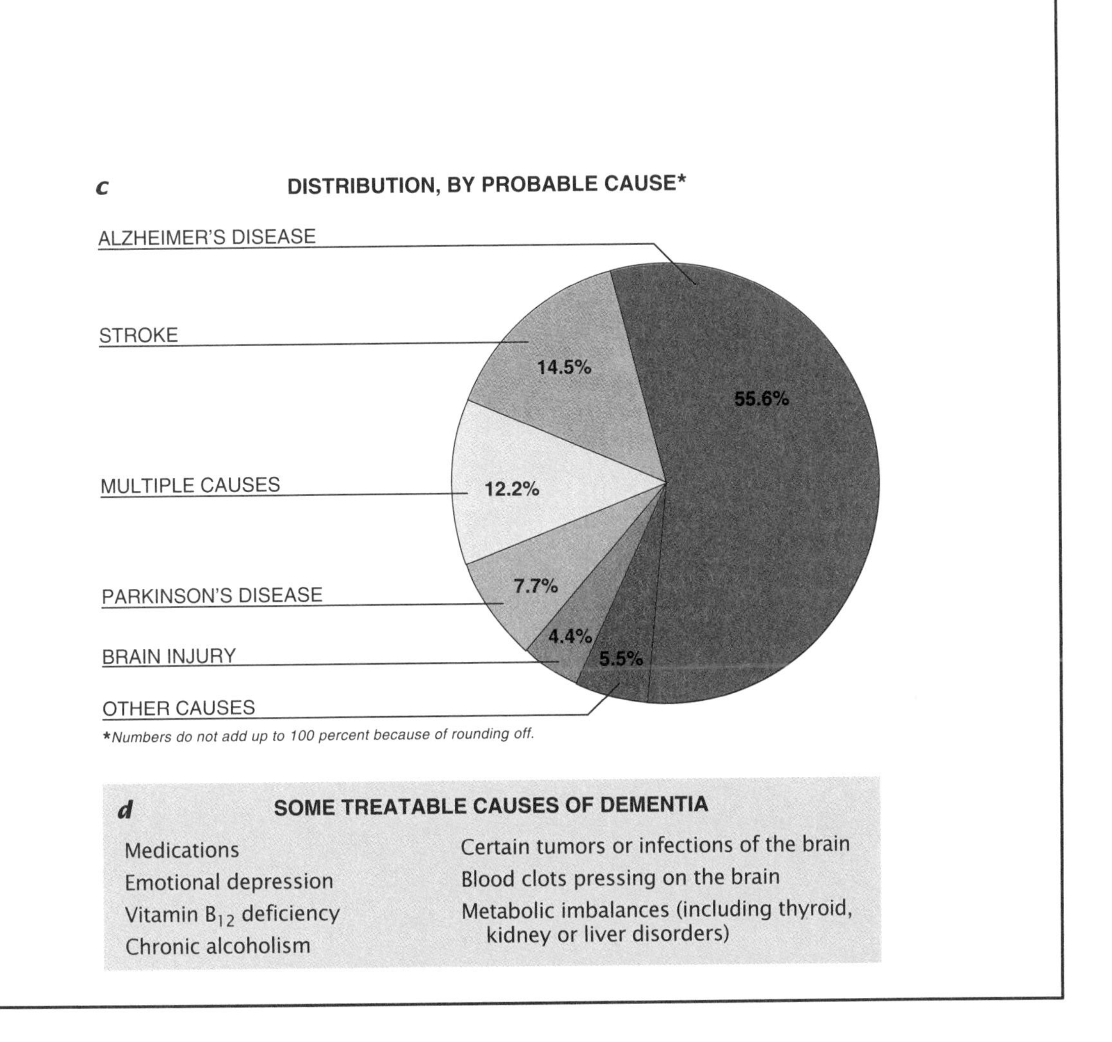

tests than do sedentary individuals of the same age with low aerobic fitness.

I would additionally advise against ingesting agents, including alcohol, that interfere with the activity of the nervous system. Similarly, I would urge physicians to be cautious when prescribing for elderly patients medications that act on the brain. Extensive experimental and clinical evidence has shown that people older than about 60 years are particularly sensitive to benzodiazepines (such as the sedative Valium) and many other depressants and stimulants of the central nervous system. Compared with young adults, older people suffer a greater decline in reasoning while such drugs are in their system, are affected longer and react to low doses more strongly. These undesirable effects on cognition are even more pronounced in those who already suffer from a dementing illness.

Researchers continue to engage in lively debate over whether maintaining or increasing mental activity can protect against cognitive decline late in life. Unfortunately, rigorous data on the subject remain elusive.

The value to brain function of dietary restriction —a well-publicized potential antidote to aging—is similarly unclear. A nutritionally balanced but very low calorie diet has been shown to delay many age-related diseases and to increase the life span in a variety of lower mammals. In some studies, rats fed restricted diets exhibited fewer neurochemical changes in their brains late in life than did their better-fed counterparts, and they were more successful in old age at learning to find their way through a maze.

Similarly, Alan Peters and his colleagues at the Boston University School of Medicine maintained rats on a very low calorie regimen that enabled them to live for as long as four years, perhaps a year longer than usual. The team found that the animals lost neurons and developed various age-related neuronal and glial alterations later in life than did control animals. On the other hand, the eventual occurrence of such alterations suggests that calorie control may delay, but will not prevent, senescence of the brain.

No one really understands the mechanism by which marked calorie restriction leads to prolonged survival in test animals. Nor does anyone know the extent to which it might retard cognitive impairment in humans. It is evident, however, that to have any benefit, the approach would probably have to be practiced throughout much of life. Moreover, sudden and severe nutritional deprivation in the elderly could lead to symptoms of dementia, making calorie restriction a risky undertaking if it is attempted without professional guidance.

A more palatable alternative (both literally and figuratively) to serious dietary denial might one day be prolonged administration of antioxidants, such as vitamin E. This vitamin has been shown to extend longevity and retard some age-associated systemic diseases in rodents, but benefits for the aging human brain have not been proved.

For now, the most rational approach to developing treatments for cognitive failure late in life is to decipher the molecular mechanisms underlying dementing diseases and then to design drugs that block one or more critical steps. In Alzheimer's disease, for example, it appears likely that therapies will ultimately be directed at inhibiting the enzymes that liberate the beta-amyloid protein from its precursor, blocking the delivery of the amyloid protein into cerebral tissue or preventing the inflammatory and neurotoxic responses that the protein apparently initiates. Such treatments might also prove to be helpful for combating mild to moderate forgetfulness in some older people who do not have full-blown dementia. This seems likely because the amyloid plaques and neurofibrillary tangles that characterize Alzheimer's disease do form in areas important for memory and learning during healthy aging, albeit to a much smaller degree. Several therapies are also under study for preventing Parkinson's disease and for preventing or treating stroke [see "Stroke Therapy," by Justin A. Zivin and Dennis W. Choi; SCIENTIFIC AMERICAN, July 1991].

During the next three decades, rigorous molecular and clinical examination of brain aging will become more common as the developed nations confront a huge surge in the numbers of very old people. An estimated three million Americans are 85 years or older today, and that number may double by the year 2020. Discovery of ways to block age-related disorders of higher cortical function without necessarily prolonging life will undoubtedly enable many of the elderly to remain independent and enjoy life well beyond the eighth decade. Successful aging of the body and mind will bring about profound economic and sociological consequences that will require great creativity and vigor to address. Fortunately, at that point, society will possess a valuable resource to help solve those problems: the sharp minds and accessible wisdom of many of its oldest citizens.

How Neural Networks Learn from Experience

Networks of artificial neurons can learn to represent complicated information. Such neural networks may provide insights into the learning abilities of the human brain.

• • •

Geoffrey E. Hinton

The brain is a remarkable computer. It interprets imprecise information from the senses at an incredibly rapid rate. It discerns a whisper in a noisy room, a face in a dimly lit alley and a hidden agenda in a political statement. Most impressive of all, the brain learns—without any explicit instructions—to create the internal representations that make these skills possible.

Much is still unknown about how the brain trains itself to process information, so theories abound. To test these hypotheses, my colleagues and I have attempted to mimic the brain's learning processes by creating networks of artificial neurons. We construct these neural networks by first trying to deduce the essential features of neurons and their interconnections. We then typically program a computer to simulate these features.

Because our knowledge of neurons is incomplete and our computing power is limited, our models are necessarily gross idealizations of real networks of neurons (see Figure 10.1). Naturally, we enthusiastically debate what features are most essential in simulating neurons. By testing these features in artificial neural networks, we have been successful at ruling out all kinds of theories about how the brain processes information. The models are also beginning to reveal how the brain may accomplish its remarkable feats of learning.

In the human brain, a typical neuron collects signals from others through a host of fine structures called dendrites. The neuron sends out spikes of electrical activity through a long, thin strand known as an axon, which splits into thousands of branches. At the end of each branch, a structure called a synapse converts the activity from the axon into electrical effects that inhibit or excite activity in the connected neurons. When a neuron receives excitatory input that is sufficiently large compared with its inhibitory input, it sends a spike of electrical activity down its axon. Learning occurs by changing the effectiveness of the synapses so that the influence of one neuron on another changes.

Artificial neural networks are typically composed of interconnected "units," which serve as model neurons. The function of the synapse is modeled by a modifiable weight, which is associated with each connection. Most artificial networks do not reflect the detailed geometry of the dendrites and axons,

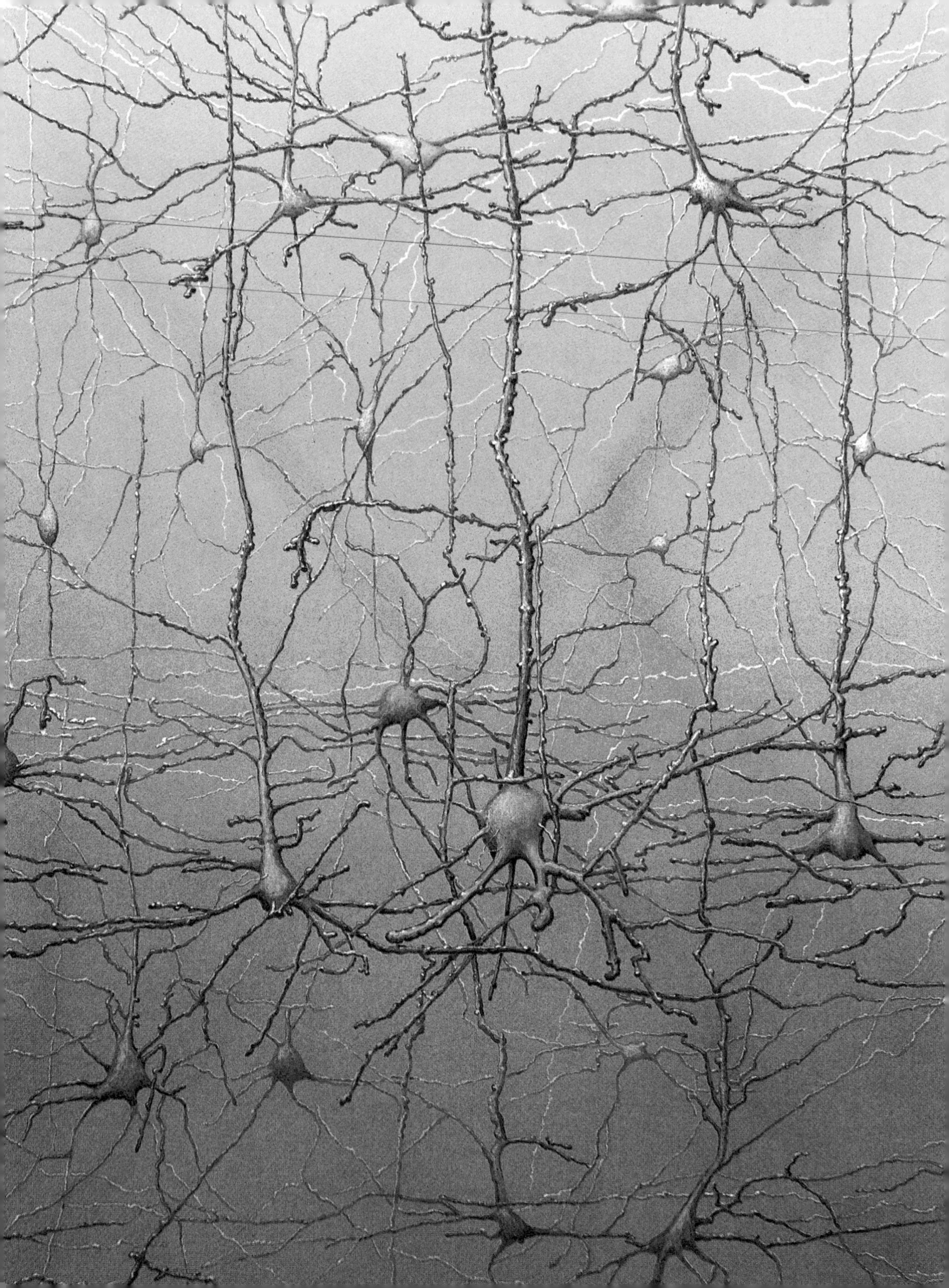

and they express the electrical output of a neuron as a single number that represents the rate of firing—its activity.

Each unit converts the pattern of incoming activities that it receives into a single outgoing activity that it broadcasts to other units. It performs this conversion in two stages. First, it multiplies each incoming activity by the weight on the connection and adds together all these weighted inputs to get a quantity called the total input. Second, a unit uses an input-output function that transforms the total input into the outgoing activity (see Figure 10.2).

The behavior of an artificial neural network depends on both the weights and the input-output function that is specified for the units. This function typically falls into one of three categories: linear, threshold or sigmoid. For linear units, the output activity is proportional to the total weighted input. For threshold units, the output is set at one of two levels, depending on whether the total input is greater than or less than some threshold value. For sigmoid units, the output varies continuously but not linearly as the input changes. Sigmoid units bear a greater resemblance to real neurons than do linear or threshold units, but all three must be considered rough approximations.

To make a neural network that performs some specific task, we must choose how the units are connected to one another, and we must set the weights on the connections appropriately. The connections determine whether it is possible for one unit to influence another. The weights specify the strength of the influence.

The most common type of artificial neural network consists of three groups, or layers, of units: a layer of input units is connected to a layer of "hidden" units, which is connected to a layer of output units (see Figure 10.3). The activity of the input units represents the raw information that is fed into the network. The activity of each hidden unit is determined by the activities of the input units and the weights on the connections between the input and hidden units. Similarly, the behavior of the output units depends on the activity of the hidden units and the weights between the hidden and output units.

This simple type of network is interesting because the hidden units are free to construct their own representations of the input. The weights between the input and hidden units determine when each hidden unit is active, and so by modifying these weights, a hidden unit can choose what it represents.

We can teach a three-layer network to perform a particular task by using the following procedure. First, we present the network with training examples, which consist of a pattern of activities for the input units together with the desired pattern of activities for the output units. We then determine how closely the actual output of the network matches the desired output. Next we change the weight of each connection so that the network produces a better approximation of the desired output.

For example, suppose we want a network to recognize handwritten digits (see boxed Figure "How a Neural Network Represents Handwritten Digits"). We might use an array of, say 256 sensors, each recording the presence or absence of ink in a small area of a single digit. The network would therefore need 256 input units (one for each sensor), 10 output units (one for each kind of digit) and a number of hidden units. For each kind of digit recorded by the sensors, the network should produce high activity in the appropriate output unit and low activity in the other output units.

To train the network, we present an image of a digit and compare the actual activity of the 10 output units with the desired activity. We then calculate the error, which is defined as the square of the difference between the actual and the desired activities. Next we change the weight of each connection so as to reduce the error. We repeat this training process for many different images of each kind of digit until the network classifies every image correctly.

To implement this procedure, we need to change each weight by an amount that is proportional to the rate at which the error changes as the weight is changed. This quantity—called the error derivative for the weight, or simply the EW—is tricky to compute efficiently. One way to calculate the EW is to perturb a weight slightly and observe how the error changes. But that method is inefficient because it requires a separate perturbation for each of the many weights.

Around 1974 Paul J. Werbos invented a much more efficient procedure for calculating the EW while he was working toward a doctorate at Harvard University. The procedure, now known as the

Figure 10.1 NETWORK OF NEURONS in the brain provides people with the ability to assimilate information. Will simulations of such networks reveal the underlying mechanisms of learning?

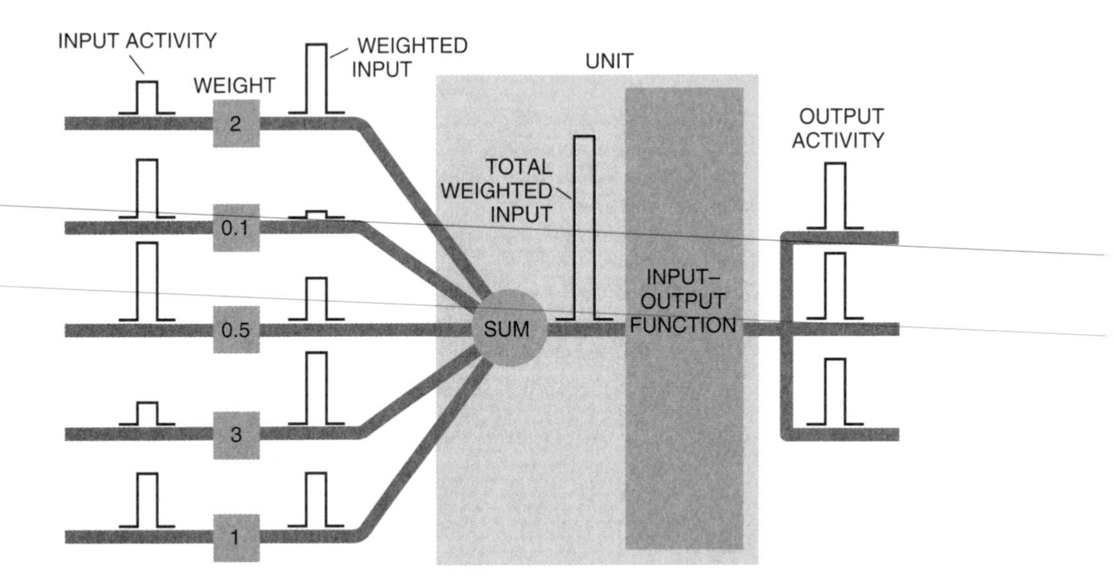

Figure 10.2 IDEALIZATION OF A NEURON processes activities, or signals. Each input activity is multiplied by a number called the weight. The "unit" adds together the weighted inputs. It then computes the output activity using an input-output function.

back-propagation algorithm, has become one of the more important tools for training neural networks (see boxed figure "The Back-Propagation Algorithm").

The back-propagation algorithm is easiest to understand if all the units in the network are linear. The algorithm computes each EW by first computing the EA, the rate at which the error changes as the activity level of a unit is changed. For output units, the EA is simply the difference between the actual and the desired output. To compute the EA for a hidden unit in the layer just before the output layer, we first identify all the weights between that hidden unit and the output units to which it is connected. We then multiply those weights by the EAs of those output units and add the products. This sum equals the EA for the chosen hidden unit. After calculating all the EAs in the hidden layer just before the output layer, we can compute in like fashion the EAs for other layers, moving from layer to layer in a direction opposite to the way activities propagate through the network. This is what gives back propagation its name. Once the EA has been computed for a unit, it is straight-forward to com-

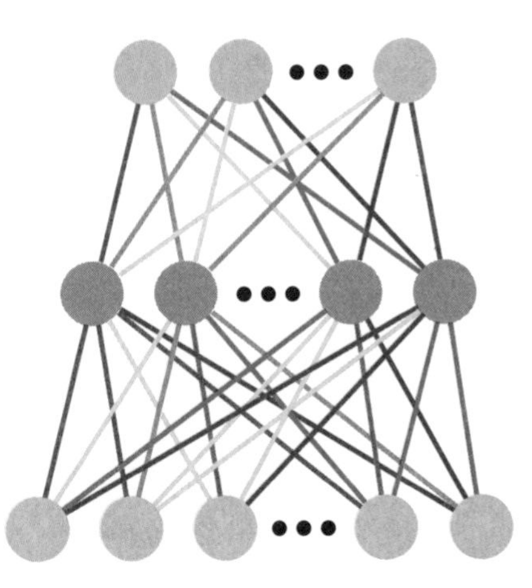

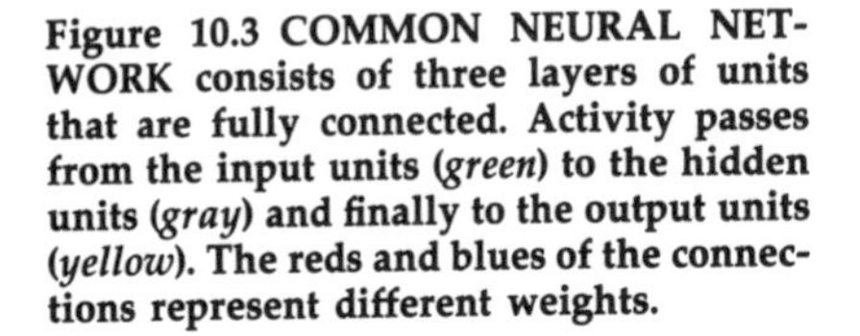
Figure 10.3 COMMON NEURAL NETWORK consists of three layers of units that are fully connected. Activity passes from the input units (*green*) to the hidden units (*gray*) and finally to the output units (*yellow*). The reds and blues of the connections represent different weights.

How a Neural Network Represents Handwritten Digits

A neural network—consisting of 256 input units, nine hidden units and 10 output units—has been trained to recognize handwritten digits. The illustration below shows the activities of the units when the network is presented with a handwritten 3. The third output unit is most active. The nine panels at the right represent the 256 incoming weights and the 10 outgoing weights for each of the nine hidden units. The red regions indicate weights that are excitatory, whereas yellow regions represent weights that are inhibitory.

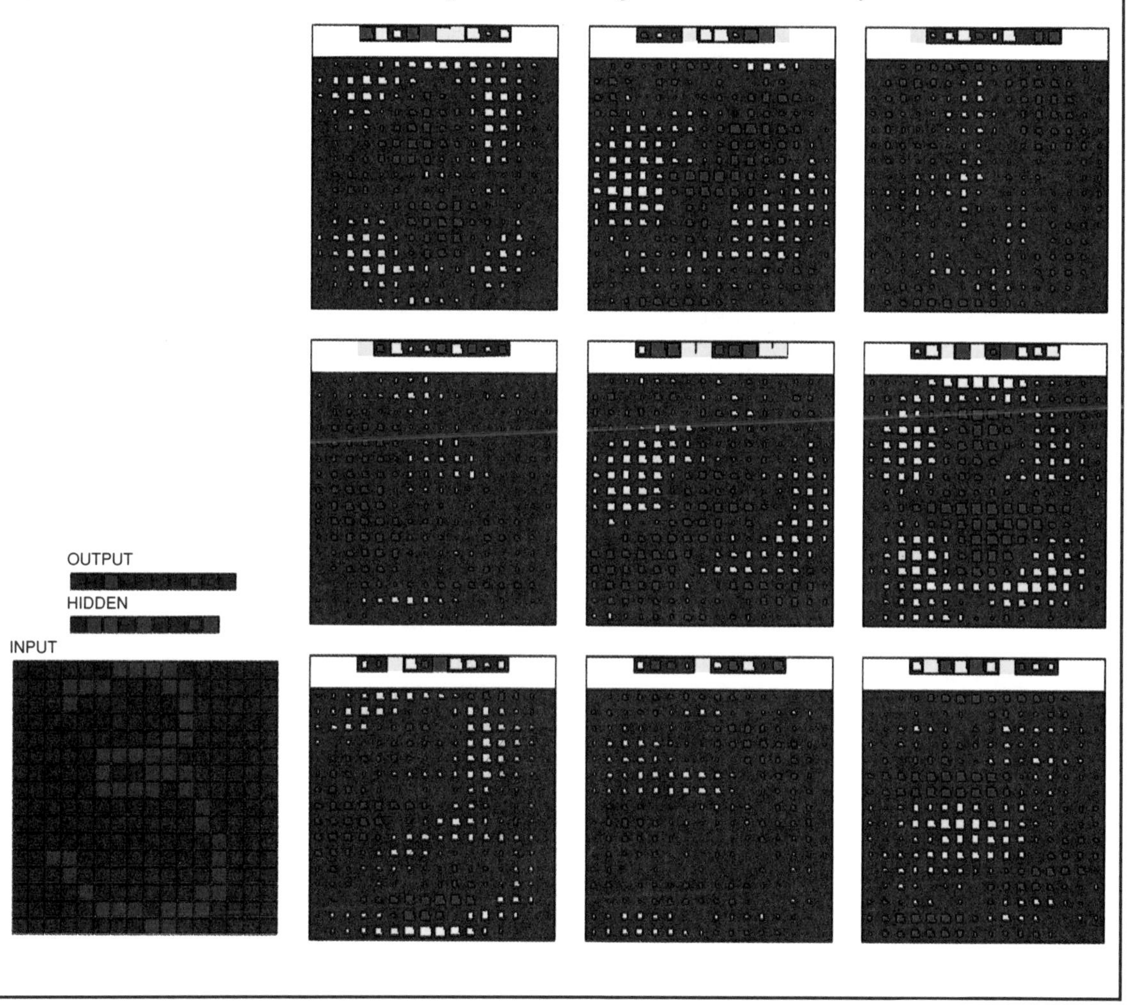

pute the EW for each incoming connection of the unit. The EW is the product of the EA and the activity through the incoming connection.

For nonlinear units, the back-propagation algorithm includes an extra step. Before back-propagating, the EA must be converted into the EI, the rate at which the error changes as the total input received by a unit is changed.

The back-propagation algorithm was largely ignored for years after its invention, probably because its usefulness was not fully appreciated. In the

The Back-Propagation Algorithm

To train a neural network to perform some task, we must adjust the weights of each unit in such a way that the error between the desired output and the actual output is reduced. This process requires that the neural network compute the error derivative of the weights (EW). In other words, it must calculate how the error changes as each weight is increased or decreased slightly. The back-propagation algorithm is the most widely used method for determining the EW.

To implement the back-propagation algorithm, we must first describe a neural network in mathematical terms. Assume that unit j is a typical unit in the output layer and unit i is a typical unit in the previous layer. A unit in the output layer determines its activity by following a two-step procedure. First, it computes the total weighted input x_j, using the formula:

$$x_j = \sum_i y_i w_{ij},$$

where y_i is the activity level of the ith unit in the previous layer and w_{ij} is the weight of the connection between the ith and jth unit.

Next, the unit calculates the activity y_j using some function of the total weighted input. Typically, we use the sigmoid function:

$$y_j = \frac{1}{1 + e^{-x_j}}.$$

Once the activities of all the output units have been determined, the network computes the error $\mathcal{E}$, which is defined by the expression:

$$\mathcal{E} = \tfrac{1}{2} \sum_j (y_j - d_j)^2,$$

where y_j is the activity level of the jth unit in the top layer and d_j is the desired output of the jth unit.

The back-propagation algorithm consists of four steps:

1. Compute how fast the error changes as the activity of an output unit is changed. This error derivative (EA) is the difference between the actual and the desired activity.

$$EA_j = \frac{\partial \mathcal{E}}{\partial y_i} = y_j - d_j$$

2. Compute how fast the error changes as the total input received by an output unit is changed. This quantity (EI) is the answer from step 1 multiplied by the rate at which the output of a unit changes as its total input is changed.

$$EI_j = \frac{\partial \mathcal{E}}{\partial x_j} = \frac{\partial \mathcal{E}}{\partial y_j}\frac{dy_j}{dx_j} = EA_j y_j (1 - y_j)$$

3. Compute how fast the error changes as a weight on the connection into an output unit is changed. This quantity (EW) is the answer from step 2 multiplied by the activity level of the unit from which the connection emanates.

$$EW_{ij} = \frac{\partial \mathcal{E}}{\partial w_{ij}} = \frac{\partial \mathcal{E}}{\partial x_j}\frac{\partial x_j}{\partial w_{ij}} = EI_j Y_i$$

4. Compute how fast the error changes as the activity of a unit in the previous layer is changed. This crucial step allows back propagation to be applied to multilayer networks. When the activity of a unit in the previous layer changes, it affects the activities of all the output units to which it is connected. So to compute the overall effect on the error, we add together all these separate effects on output units. But each effect is simple to calculate. It is the answer in step 2 multiplied by the weight on the connection to that output unit.

$$EA_i = \frac{\partial \mathcal{E}}{\partial y_i} = \sum_j \frac{\partial \mathcal{E}}{\partial x_j}\frac{\partial x_j}{\partial y_i} = \sum_j EI_j w_{ij}$$

By using steps 2 and 4, we can convert the EAs of one layer of units into EAs for the previous layer. This procedure can be repeated to get the EAs for as many previous layers as desired. Once we know the EA of a unit, we can use steps 2 and 3 to compute the EWs on its incoming connections.

early 1980s David E. Rumelhart, then at the University of California at San Diego, and David B. Parker, then at Stanford University, independently rediscovered the algorithm. In 1986 Rumelhart, Ronald J. Williams, also at the University of California at San Diego, and I popularized the algorithm by demonstrating that it could teach the hidden units to produce interesting representations of complex input patterns.

The back-propagation algorithm has proved surprisingly good at training networks with multiple layers to perform a wide variety of tasks. It is most useful in situations in which the relation between input and output is nonlinear and training data are abundant. By applying the algorithm, researchers have produced neural networks that recognize handwritten digits, predict currency exchange rates and maximize the yields of chemical processes. They have even used the algorithm to train networks that identify precancerous cells in Pap smears and that adjust the mirror of a telescope so as to cancel out atmospheric distortions.

Within the field of neuroscience, Richard Andersen of the Massachusetts Institute of Technology and David Zipser of the University of California at San Diego showed that the back-propagation algorithm is a useful tool for explaining the function of some neurons in the brain's cortex. They trained a neural network to respond to visual stimuli using back propagation. They then found that the responses of the hidden units were remarkably similar to those of real neurons responsible for converting visual information from the retina into a form suitable for deeper visual areas of the brain.

Yet back propagation has had a rather mixed reception as a theory of how biological neurons learn. On the one hand, the back-propagation algorithm has made a valuable contribution at an abstract level. The algorithm is quite good at creating sensible representations in the hidden units. As a result, researchers gained confidence in learning procedures in which weights are gradually adjusted to reduce errors. Previously, many workers had assumed that such methods would be hopeless because they would inevitably lead to locally optimal but globally terrible solutions. For example, a digit-recognition network might consistently home in on a set of weights that makes the network confuse ones and sevens even though an ideal set of weights exists that would allow the network to discriminate between the digits. This fear supported a widespread belief that a learning procedure was interesting only if it were guaranteed to converge eventually on the globally optimal solution. Back propagation showed that for many tasks global convergence was not necessary to achieve good performance.

On the other hand, back propagation seems biologically implausible. The most obvious difficulty is that information must travel through the same connections in the reverse direction, from one layer to the previous layer. Clearly, this does not happen in real neurons. But this objection is actually rather superficial. The brain has many pathways from later layers back to earlier ones, and it could use these pathways in many ways to convey the information required for learning.

A more important problem is the speed of the back-propagation algorithm. Here the central issue is how the time required to learn increases as the network gets larger. The time taken to calculate the error derivatives for the weights on a given training example is proportional to the size of the network because the amount of computation is proportional to the number of weights. But bigger networks typically require more training examples, and they must update the weights more times. Hence, the learning time grows much faster than does the size of the network.

The most serious objection to back propagation as a model of real learning is that it requires a teacher to supply the desired output for each training example. In contrast, people learn most things without the help of a teacher. Nobody presents us with a detailed description of the internal representations of the world that we must learn to extract from our sensory input. We learn to understand sentences or visual scenes without any direct instructions.

How can a network learn appropriate internal representations if it starts with no knowledge and no teacher? If a network is presented with a large set of patterns but is given no information about what to do with them, it apparently does not have a well-defined problem to solve. Nevertheless, researchers have developed several general-purpose, unsupervised procedures that can adjust the weights in the network appropriately.

All these procedures share two characteristics: they appeal, implicitly or explicitly, to some notion of the quality of a representation, and they work by changing the weights to improve the quality of the representation extracted by the hidden units.

In general, a good representation is one that can be described very economically but nonetheless contains enough information to allow a close approximation of the raw input to be reconstructed. For example, consider an image consisting of several ellipses. Suppose a device translates the image into an array of a million tiny squares, each of which is either light or dark. The image could be represented simply by the positions of the dark squares. But other, more efficient representations are also possible. Ellipses differ in only five ways: orientation, vertical position, horizontal position, length and width. The image can therefore be described using only five parameters per ellipse (see Figure 10.4).

Although describing an ellipse by five parameters requires more bits than describing a single dark square by two coordinates, we get an overall savings because far fewer parameters than coordinates are needed. Furthermore, we do not lose any information by describing the ellipses in terms of their parameters: given the parameters of the ellipse, we could reconstruct the original image if we so desired.

Almost all the unsupervised learning procedures can be viewed as methods of minimizing the sum of two terms, a code cost and a reconstruction cost. The code cost is the number of bits required to describe the activities of the hidden units. The reconstruction cost is the number of bits required to describe the misfit between the raw input and the best approximation to it that could be reconstructed from the activities of the hidden units. The reconstruction cost is proportional to the squared difference between the raw input and its reconstruction.

Two simple methods for discovering economical codes allow fairly accurate reconstruction of the input: principal-components learning and competitive learning. In both approaches, we first decide how economical the code should be and then modify the weights in the network to minimize the reconstruction error.

A principal-components learning strategy is based on the idea that if the activities of pairs of input units are correlated in some way, it is a waste of bits to describe each input activity separately. A more efficient approach is to extract and describe the principal components—that is, the components of variation shared by many input units. If we wish to discover, say, 10 of the principal components, then we need only a single layer of 10 hidden units.

Because such networks represent the input using only a small number of components, the code cost is low. And because the input can be reconstructed quite well from the principal components, the reconstruction cost is small.

One way to train this type of network is to force it to reconstruct an approximation to the input on a set of output units. Then back propagation can be used to minimize the difference between the actual output and the desired output. This process resembles supervised learning, but because the desired output is exactly the same as the input, no teacher is required.

Many researchers, including Ralph Linsker of the IBM Thomas J. Watson Research Center and Erkki Oja of Lappeenranta University of Technology in Finland, have discovered alternative algorithms for learning principal components. These algorithms

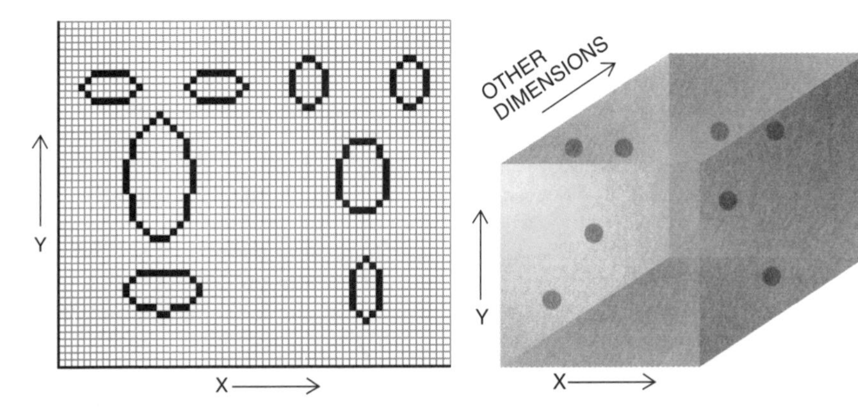

Figure 10.4 TWO FACES composed of eight ellipses can be represented as many points in two dimensions. Alternatively, because the ellipses differ in only five ways—orientation, vertical position, horizontal position, length and width—the two faces can be represented as eight points in a five-dimensional space.

are more biologically plausible because they do not require output units or back propagation. Instead they use the correlation between the activity of a hidden unit and the activity of an input unit to determine the change in the weight.

When a neural network uses principal-components learning, a small number of hidden units cooperate in representing the input pattern. In contrast, in competitive learning, a large number of hidden units compete so that a single hidden unit is used to represent any particular input pattern. The selected hidden unit is the one whose incoming weights are most similar to the input pattern (see Figure 10.5).

Now suppose we had to reconstruct the input pattern solely from our knowledge of which hidden unit was chosen. Our best bet would be to copy the pattern of incoming weights of the chosen hidden unit. To minimize the reconstruction error, we should move the pattern of weights of the winning hidden unit even closer to the input pattern. This is what competitive learning does. If the network is presented with training data that can be grouped into clusters of similar input patterns, each hidden unit learns to represent a different cluster, and its incoming weights converge on the center of the cluster.

Like the principal-components algorithm, competitive learning minimizes the reconstruction cost while keeping the code cost low. We can afford to use many hidden units because even with a million units it takes only 20 bits to say which one won.

In the early 1980s Teuvo Kohonen of Helsinki University introduced an important modification of the competitive learning algorithm. Kohonen showed how to make physically adjacent hidden units learn to represent similar input patterns. Kohonen's algorithm adapts not only the weights of the winning hidden unit but also the weights of the winners' neighbors. The algorithm's ability to map

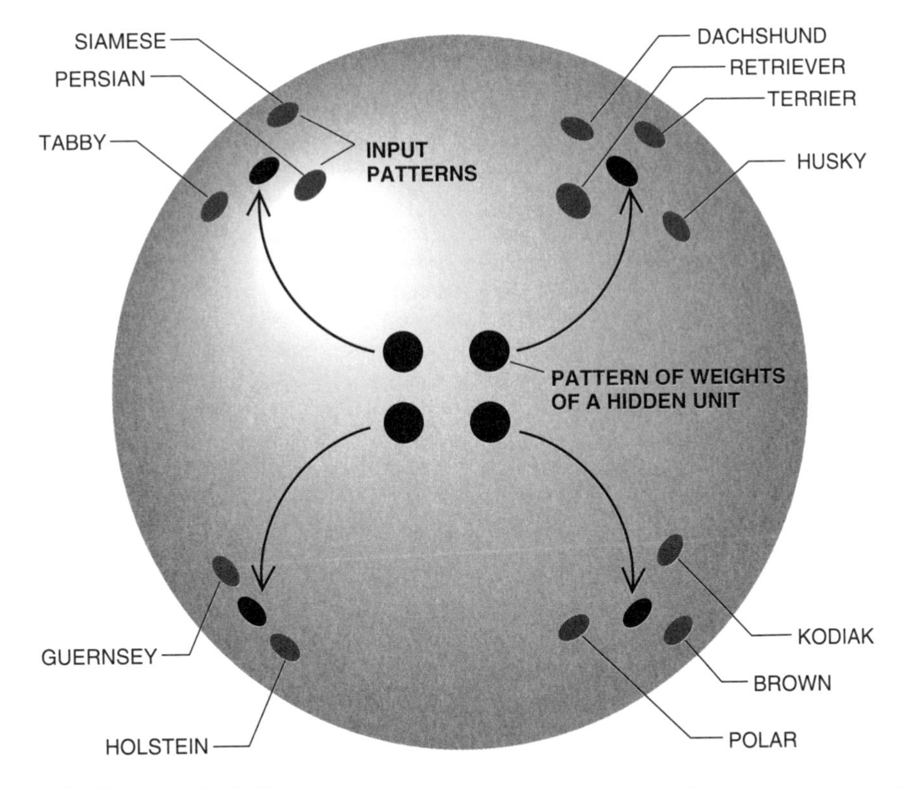

Figure 10.5 COMPETITIVE LEARNING can be envisioned as a process in which each input pattern attracts the weight pattern of the closest hidden unit. Each input pattern represents a set of distinguishing features. The weight patterns of hidden units are adjusted so that they migrate slowly toward the closest set of input patterns. In this way, each hidden unit learns to represent a cluster of similar input patterns.

similar input patterns to nearby hidden units suggests that a procedure of this type may be what the brain uses to create the topographic maps found in the visual cortex (see Chapter 3, "The Visual Image in Mind and Brain," by Semir Zeki).

Unsupervised learning algorithms can be classified according to the type of representation they create. In principal-components methods, the hidden units cooperate, and the representation of each input pattern is distributed across all of them. In competitive methods, the hidden units compete, and the representation of the input pattern is localized in the single hidden unit that is selected. Until recently, most work on unsupervised learning focused on one or another of these two techniques, probably because they lead to simple rules for changing the weights. But the most interesting and powerful algorithms probably lie somewhere between the extremes of purely distributed and purely localized representations.

Horace B. Barlow of the University of Cambridge has proposed a model in which each hidden unit is rarely active and the representation of each input pattern is distributed across a small number of selected hidden units. He and his co-workers have shown that this type of code can be learned by forcing hidden units to be uncorrelated while also ensuring that the hidden code allows good reconstruction of the input.

Unfortunately, most current methods of minimizing the code cost tend to eliminate all the redundancy among the activities of the hidden units. As a result, the network is very sensitive to the malfunction of a single hidden unit. This feature is uncharacteristic of the brain, which is generally not affected greatly by the loss of a few neurons.

The brain seems to use what are known as population codes, in which information is represented by a whole population of active neurons. That point was beautifully demonstrated in the experiments of David L. Sparks and his co-workers at the University of Alabama. While investigating how the brain of a monkey instructs its eyes where to move, they found that the required movement is encoded by the activities of a whole population of cells, each of which represents a somewhat different movement. The eye movement that is actually made corresponds to the average of all the movements encoded by the active cells. If some brain cells are anesthetized, the eye moves to the point associated with the average of the remaining active cells. Population codes may be used to encode not only eye movements but also faces, as shown by Malcolm P. Young and Shigeru Yamane at the RIKEN Institute in Japan in recent experiments on the inferior temporal cortex of monkeys.

For both eye movements and faces, the brain must represent entities that vary along many different dimensions. In the case of an eye movement, there are just two dimensions, but for something like a face, there are dimensions such as happiness, hairiness or familiarity, as well as spatial parameters such as position, size and orientation. If we associate with each face-sensitive cell the parameters of the face that make it most active, we can average these parameters over a population of active cells to discover the parameters of the face being represented by that population code. In abstract terms, each face cell represents a particular point in a multidimensional space of possible faces, and any face can then be represented by activating all the cells that encode very similar faces, so that a bump of activity appears in the multidimensional space of possible faces (see Figure 10.6).

Population coding is attractive because it works even if some of the neurons are damaged. It can do so because the loss of a random subset of neurons has little effect on the population average. The same reasoning applies if some neurons are overlooked when the system is in a hurry. Neurons communicate by sending discrete spikes called action potentials, and in a very short time interval many of the "active" neurons may not have time to send a spike. Nevertheless, even in such a short interval, a population code in one part of the brain can still give rise to an approximately correct population code in another part of the brain.

At first sight, the redundancy in population codes seems incompatible with the idea of constructing internal representations that minimize the code cost. Fortunately, we can overcome this difficulty by using a less direct measure of code cost. If the activity that encodes a particular entity is a smooth bump in which activity falls off in a standard way as we move away from the center, we can describe the bump of activity completely merely by specifying its center. So a fairer measure of code cost is the cost of describing the center of the bump of activity plus the cost of describing how the actual activities of the units depart from the desired smooth bump of activity.

Using this measure of the code cost, we find that population codes are a convenient way of extracting

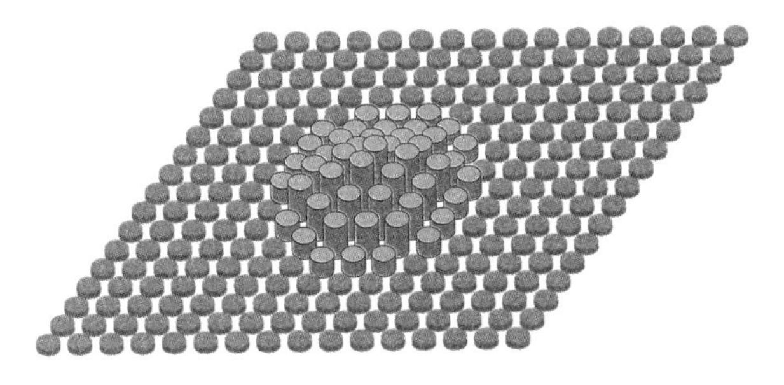

Figure 10.6 POPULATION CODING represents a multiparameter object as a bump of activity spread over many hidden units. Each disk represents an inactive hidden unit. Each cylinder indicates an active unit, and its height depicts the level of activity.

a hierarchy of progressively more efficient encodings of the sensory input. This point is best illustrated by a simple example. Consider a neural network that is presented with an image of a face (see Figure 10.7). Suppose the network already contains one set of units dedicated to representing noses, another set for mouths and another set for eyes. When it is shown a particular face, there will be one bump of activity in the nose units, one in the mouth units and two in the eye units. The location of each of these activity bumps represents the spatial parameters of the feature encoded by the bump. Describing the four activity bumps is cheaper than describing the raw image, but it would obviously be cheaper still to describe a single bump of activity in a set of face units, assuming of course that the nose, mouth and eyes are in the correct spatial relations to form a face.

This raises an interesting issue: How can the network check that the parts are correctly related to one another to make a face? Some time ago Dana H. Ballard of the University of Rochester introduced a clever technique for solving this type of problem that works nicely with population codes.

If we know the position, size and orientation of a nose, we can predict the position, size and orientation of the face to which it belongs because the spatial relation between noses and faces is roughly fixed. We therefore set the weights in the neural network so that a bump of activity in the nose units tries to cause an appropriately related bump of activity in the face units. But we also set the thresholds of the face units so that the nose units alone are insufficient to activate the face units. If, however, the bump of activity in the mouth units also tries to cause a bump in the same place in the face units, then the thresholds can be overcome. In effect, we have checked that the nose and mouth are correctly related to each other by checking that they both predict the same spatial parameters for the whole face.

This method of checking spatial relations is intriguing because it makes use of the kind of redundancy between different parts of an image that unsupervised learning should be good at finding. It therefore seems natural to try to use unsupervised learning to discover hierarchical population codes for extracting complex shapes. In 1986 Eric Saund of M.I.T. demonstrated one method of learning simple population codes for shapes. It seems likely that with a clear definition of the code cost, an unsupervised network will be able to discover more complex hierarchies by trying to minimize the cost of coding the image. Richard Zemel and I at the University of Toronto are now investigating this possibility.

By using unsupervised learning to extract a hierarchy of successively more economical representations, it should be possible to improve greatly the speed of learning in large multilayer networks. Each layer of the network adapts its incoming weights to make its representation better than the representation in the previous layer, so weights in one layer can be learned without reference to weights in subsequent layers. This strategy eliminates many of the interactions between weights that make backpropagation learning very slow in deep multilayer networks.

All the learning procedures discussed thus far are implemented in neural networks in which activity flows only in the forward direction from input to output even though error derivatives may flow in the backward direction. Another important possibility to consider is networks in which activity flows around closed loops. Such recurrent networks may settle down to stable states, or they may exhibit complex temporal dynamics that can be used to produce sequential behavior. If they settle to stable states, error derivatives can be computed using methods much simpler than back propagation.

Although investigators have devised some powerful learning algorithms that are of great practical value, we still do not know which representations

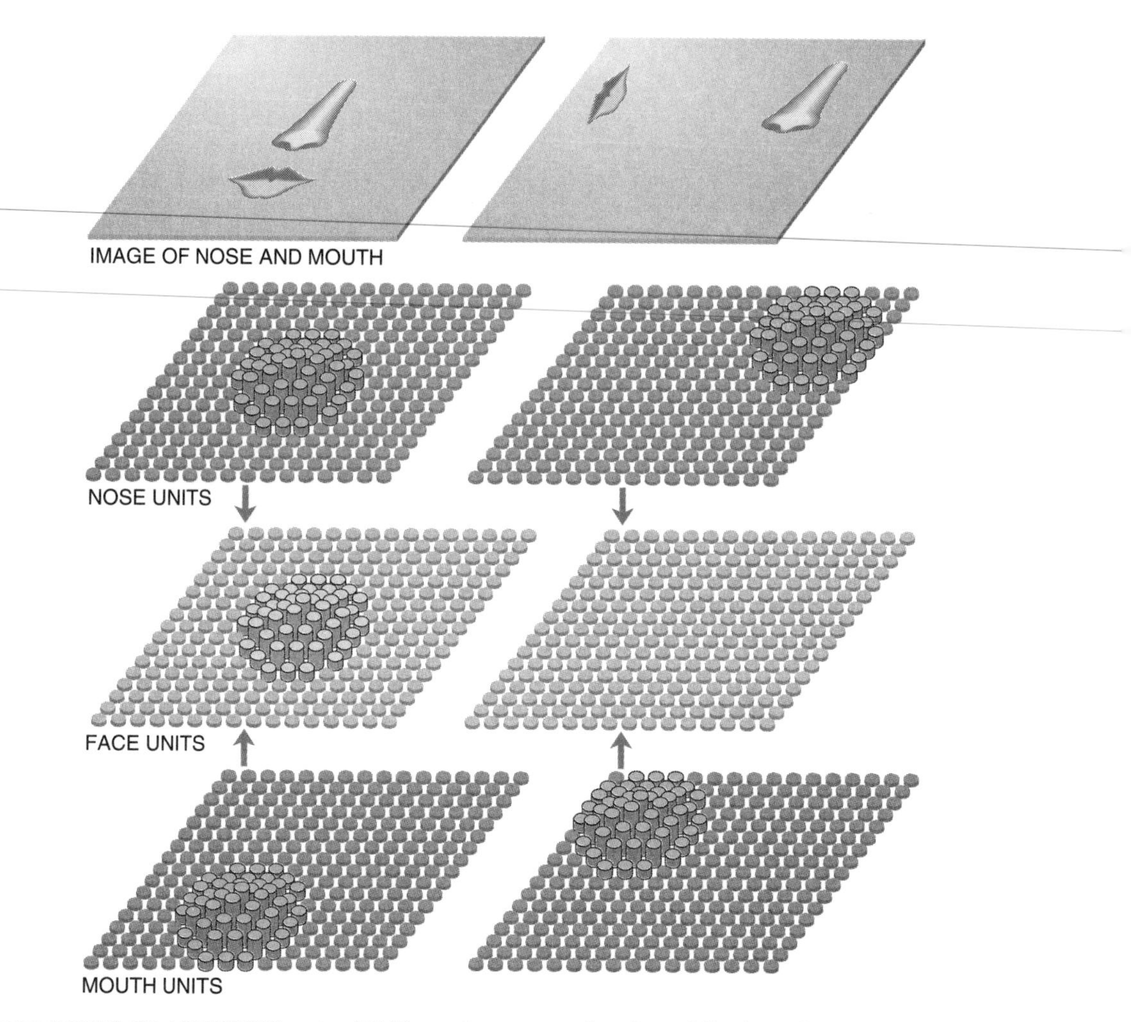

Figure 10.7 BUMPS OF ACTIVITY in sets of hidden units represent the image of a nose and a mouth. These population codes will cause a bump in the face units if the nose and mouth have the correct spatial relation (*left*). If not, the active nose units will try to create a bump in the face units at one location while the active mouth units will do the same at a different location. As a result, the input activity to the face units does not exceed a threshold value, and no bump is formed in the face units (*right*).

and learning procedures are actually used by the brain. But sooner or later computational studies of learning in artificial neural networks will converge on the methods discovered by evolution. When that happens, a lot of diverse empirical data about the brain will suddenly make sense, and many new applications of artificial neural networks will become feasible.

The Problem of Consciousness

It can now be approached by scientific investigation of the visual system. The solution will require a close collaboration among psychologists, neuroscientists and theorists.

• • •

Francis Crick and Christof Koch

The overwhelming question in neurobiology today is the relation between the mind and the brain. Everyone agrees that what we know as mind is closely related to certain aspects of the behavior of the brain, not to the heart, as Aristotle thought. Its most mysterious aspect is consciousness or awareness, which can take many forms, from the experience of pain to self-consciousness.

In the past the mind (or soul) was often regarded, as it was by Descartes, as something immaterial, separate from the brain but interacting with it in some way. A few neuroscientists, such as Sir John Eccles, still assert that the soul is distinct from the body. But most neuroscientists now believe that all aspects of mind, including its most puzzling attribute—consciousness or awareness—are likely to be explainable in a more materialistic way as the behavior of large sets of interacting neurons. As William James, the father of American psychology, said a century ago, consciousness is not a thing but a process.

Exactly what the process is, however, has yet to be discovered. For many years after James penned *The Principles of Psychology*, consciousness was a taboo concept in American psychology because of the dominance of the behaviorist movement. With the advent of cognitive science in the mid-1950s, it became possible once more for psychologists to consider mental processes as opposed to merely observing behavior. In spite of these changes, until recently most cognitive scientists ignored consciousness, as did almost all neuroscientists. The problem was felt to be either purely "philosophical" or too elusive to study experimentally. It would not have been easy for a neuroscientist to get a grant just to study consciousness.

In our opinion, such timidity is ridiculous, so a few years ago we began to think about how best to attack the problem scientifically. How to explain mental events as being caused by the firing of large sets of neurons? Although there are those who believe such an approach is hopeless, we feel it is not productive to worry too much over aspects of the problem that cannot be solved scientifically or, more precisely, cannot be solved solely by using existing scientific ideas. Radically new concepts may indeed be needed—recall the modifications of scientific thinking forced on us by quantum mechanics. The only sensible approach is to press the experimental attack until we are confronted with dilemmas that call for new ways of thinking.

There are many possible approaches to the problem of consciousness. Some psychologists feel that

any satisfactory theory should try to explain as many aspects of consciousness as possible, including emotion, imagination, dreams, mystical experiences and so on. Although such an all-embracing theory will be necessary in the long run, we thought it wiser to begin with the particular aspect of consciousness that is likely to yield most easily. What this aspect may be is a matter of personal judgment. We selected the mammalian visual system because humans are very visual animals and because so much experimental and theoretical work has already been done on it (see Chapter 3, "The Visual Image of Mind and Brain," by Semir Zeki).

It is not easy to grasp exactly what we need to explain, and it will take many careful experiments before visual consciousness can be described scientifically. We did not attempt to define consciousness itself because of the dangers of premature definition. (If this seems like a cop-out, try defining the word "gene"—you will not find it easy.) Yet the experimental evidence that already exists provides enough of a glimpse of the nature of visual consciousness to guide research. In this chapter, we will attempt to show how this evidence opens the way to attack this profound and intriguing problem.

Visual theorists agree that the problem of visual consciousness is ill posed. The mathematical term "ill posed" means that additional constraints are needed to solve the problem. Although the main function of the visual system is to perceive objects and events in the world around us, the information available to our eyes is not sufficient by itself to provide the brain with its unique interpretation of the visual world. The brain must use past experience (either its own or that of our distant ancestors, which is embedded in our genes) to help interpret the information coming into our eyes. An example would be the derivation of the three-dimensional representation of the world from the two-dimensional signals falling onto the retinas of our two eyes or even onto one of them.

Visual theorists also would agree that seeing is a constructive process, one in which the brain has to carry out complex activities (sometimes called computations) in order to decide which interpretation to adopt of the ambiguous visual input. "Computation" implies that the brain acts to form a symbolic representation of the visual world, with a mapping (in the mathematical sense) of certain aspects of that world onto elements in the brain.

Ray Jackendoff of Brandeis University postulates, as do most cognitive scientists, that the computations carried out by the brain are largely unconscious and that what we become aware of is the result of these computations. But while the customary view is that this awareness occurs at the highest levels of the computational system, Jackendoff has proposed an intermediate-level theory of consciousness.

What we see, Jackendoff suggests, relates to a representation of surfaces that are directly visible to us, together with their outline, orientation, color, texture and movement. (This idea has similarities to what the late David C. Marr of the Massachusetts Institute of Technology called a "2½-dimensional sketch." It is more than a two-dimensional sketch because it conveys the orientation of the visible surfaces. It is less than three-dimensional because depth information is not explicitly represented.) In the next stage this sketch is processed by the brain to produce a three-dimensional representation. Jackendoff argues that we are not visually aware of this three-dimensional representation.

An example may make this process clearer. If you look at a person whose back is turned to you, you can see the back of the head but not the face. Nevertheless, your brain infers that the person has a face. We can deduce as much because if that person turned around and had no face, you would be very surprised (see Figure 11.1).

The viewer-centered representation that corresponds to the visible back of the head is what you are vividly aware of. What your brain infers about the front would come from some kind of three-dimensional representation. This does not mean that information flows only from the surface representation to the three-dimensional one; it almost certainly flows in both directions. When you imagine the front of the face, what you are aware of is a surface representation generated by information from the three-dimensional model.

It is important to distinguish between an explicit and an implicit representation. An explicit repre-

Figure 11.1 VISUAL AWARENESS primarily involves seeing what is directly in front of you, but it can be influenced by a three-dimensional representation of the object in view retained by the brain. If you see the back of a person's head, the brain infers that there is a face on the front of it. We know this is true because we would be very startled if a mirror revealed that the front was exactly like the back, as in this painting, *Reproduction Prohibited* (1937), by René Magritte.

sentation is something that is symbolized without further processing. An implicit representation contains the same information but requires further processing to make it explicit. The pattern of colored dots on a television screen, for example, contains an implicit representation of objects (say, a person's face), but only the dots and their locations are explicit. When you see a face on the screen, there must be neurons in your brain whose firing, in some sense, symbolizes that face.

We call this pattern of firing neurons an active representation. A latent representation of a face must also be stored in the brain, probably as a special pattern of synaptic connections between neurons (see Chapter 10, "How Neural Networks Learn from Experience," by Geoffrey E. Hinton). For example, you probably have a representation of the Statue of Liberty in your brain, a representation that usually is inactive. If you do think about the Statue, the representation becomes active, with the relevant neurons firing away.

An object, incidentally, may be represented in more than one way—as a visual image, as a set of words and their related sounds, or even as a touch or a smell. These different representations are likely to interact with one another. The representation is likely to be distributed over many neurons, both locally, as discussed in Geoffrey E. Hinton's chapter, and more globally. Such a representation may not be as simple and straightforward as uncritical introspection might indicate. There is suggestive evidence, partly from studying how neurons fire in various parts of a monkey's brain and partly from examining the effects of certain types of brain damage in humans, that different aspects of a face—and of the implications of a face—may be represented in different parts of the brain.

First, there is the representation of a face as a face: two eyes, a nose, a mouth and so on. The neurons involved are usually not too fussy about the exact size or position of this face in the visual field, nor are they very sensitive to small changes in its orientation. In monkeys, there are neurons that respond best when the face is turning in a particular direction, while others seem to be more concerned with the direction in which the eyes are gazing.

Then there are representations of the parts of a face, as separate from those for the face as a whole. Further, the implications of seeing a face, such as that person's sex, the facial expression, the familiarity or unfamiliarity of the face, and in particular whose face it is, may each be correlated with neurons firing in other places.

What we are aware of at any moment, in one sense or another, is not a simple matter. We have suggested that there may be a very transient form of fleeting awareness that represents only rather simple features and does not require an attentional mechanism. From this brief awareness the brain constructs a viewer-centered representation—what we see vividly and clearly—that does require attention. This in turn probably leads to three-dimensional object representations and thence to more cognitive ones.

Representations corresponding to vivid consciousness are likely to have special properties. William James (see Figure 11.2) thought that consciousness involved both attention and short-term memory. Most psychologists today would agree with this view. Jackendoff writes that consciousness is "enriched" by attention, implying that while attention may not be essential for certain limited types of consciousness, it is necessary for full consciousness.

Yet it is not clear exactly which forms of memory are involved. Is long-term memory needed? Some forms of acquired knowledge are so embedded in the machinery of neural processing that they are almost certainly used in becoming aware of something. On the other hand, there is evidence from studies of brain-damaged patients (such as H.M., described in Chapter 4, "The Biological Basis of Learning and Individuality," by Eric R. Kandel and Robert D. Hawkins) that the ability to lay down new long-term episodic memories is not essential for consciousness.

It is difficult to imagine that anyone could be conscious if he or she had no memory whatsoever of what had just happened, even an extremely short one. Visual psychologists talk of iconic memory, which lasts for a fraction of a second, and working memory (such as that used to remember a new telephone number) that lasts for only a few seconds unless it is rehearsed. It is not clear whether both of these are essential for consciousness. In any case, the division of short-term memory into these two categories may be too crude.

If these complex processes of visual awareness are localized in parts of the brain, which processes are likely to be where? Many regions of the brain may be involved, but it is almost certain that the cerebral neocortex plays a dominant role. Visual

Figure 11.2 WILLIAM JAMES, the father of American psychology, observed that consciousness is not a thing but a process.

information from the retina reaches the neocortex mainly by way of a part of the thalamus (the lateral geniculate nucleus); another significant visual pathway from the retina is to the superior colliculus, at top of the brain stem.

The cortex in humans consists of two intricately folded sheets of nerve tissue, one on each side of the head. These sheets are connected by a large tract of about half a billion axons called the corpus callosum. It is well known that if the corpus callosum is cut, as is done for certain cases of intractable epilepsy, one side of the brain is not aware of what the other side is seeing.

In particular, the left side of the brain (in a right-handed person) appears not to be aware of visual information received exclusively by the right side. This shows that none of the information required for visual awareness can reach the other side of the brain by traveling down to the brain stem and, from there, back up. In a normal person, such information can get to the other side only by using the axons in the corpus callosum.

A different part of the brain—the hippocampal system—is involved in one-shot, or episodic, memories that, over weeks and months, it passes on to the neocortex, as described in Chapter 4 by Eric R. Kandel and Robert D. Hawkins. This system is so placed that it receives inputs from, and projects to, many parts of the brain.

Thus, one might suspect that the hippocampal system is the essential seat of consciousness. This is not the case: evidence from studies of patients with damaged brains shows that this system is not essential for visual awareness, although naturally a patient lacking one, such as H.M., is severely handicapped in everyday life because he cannot remember anything that took place more than a minute or so in the past.

In broad terms, the neocortex of alert animals probably acts in two ways. By building on crude and somewhat redundant wiring, produced by our genes and by embryonic processes (see Chapter 2, "The Developing Brain," by Carla J. Shatz), the neocortex draws on visual and other experience to slowly "rewire" itself to create categories (or "features") it can respond to. A new category is not fully created in the neocortex after exposure to only one example of it, although some small modifications of the neural connections may be made.

The second function of the neocortex (at least of

the visual part of it) is to respond extremely rapidly to incoming signals. To do so, it uses the categories it has learned and tries to find the combinations of active neurons that, on the basis of its past experience, are most likely to represent the relevant objects and events in the visual world at that moment. The formation of such coalitions of active neurons may also be influenced by biases coming from other parts of the brain: for example, signals telling it what best to attend to or high-level expectations about the nature of the stimulus.

Consciousness, as James noted, is always changing. These rapidly formed coalitions occur at different levels and interact to form even broader coalitions. They are transient, lasting usually for only a fraction of a second. Because coalitions in the visual system are the basis of what we see, evolution has seen to it that they form as fast as possible; otherwise, no animal could survive. The brain is handicapped in forming neuronal coalitions rapidly because, by computer standards, neurons act very slowly. The brain compensates for this relative slowness partly by using very many neurons, simultaneously and in parallel, and partly by arranging the system in a roughly hierarchical manner.

If visual awareness at any moment corresponds to sets of neurons firing, then the obvious question is: Where are these neurons located in the brain, and in what way are they firing? Visual awareness is highly unlikely to occupy all the neurons in the neocortex that happen to be firing above their background rate at a particular moment. We would expect that, theoretically, at least some of these neurons would be involved in doing computations —trying to arrive at the best coalitions—while others would express the results of these computations, in other words, what we see.

Fortunately, some experimental evidence can be found to back up this theoretical conclusion. A phenomenon called binocular rivalry may help identify the neurons whose firing symbolizes awareness. This phenomenon can be seen in dramatic form in an exhibit prepared by Sally Duensing and Bob Miller at the Exploratorium in San Francisco (see boxed figure "The Cheshire Cat Experiment).

Binocular rivalry occurs when each eye has a different visual input relating to the same part of the visual field. The early visual system on the left side of the brain receives an input from both eyes but sees only the part of the visual field to the right of the fixation point. The converse is true for the right side. If these two conflicting inputs are rivalrous, one sees not the two inputs superimposed but first one input, then the other, and so on in alternation.

In the exhibit, called "The Cheshire Cat," viewers put their heads in a fixed place and are told to keep the gaze fixed. By means of a suitably placed mirror, one of the eyes can look at another person's face, directly in front, while the other eye sees a blank white screen to the side. If the viewer waves a hand in front of this plain screen at the same location in his or her visual field occupied by the face, the face is wiped out. The movement of the hand, being visually very salient, has captured the brain's attention. Without attention the face cannot be seen. If the viewer moves the eyes, the face reappears.

In some cases, only part of the face disappears. Sometimes, for example, one eye, or both eyes, will remain. If the viewer looks at the smile on the person's face, the face may disappear, leaving only the smile. For this reason, the effect has been called the Cheshire Cat effect, after the cat in Lewis Carroll's *Alice's Adventures in Wonderland.*

Although it is very difficult to record activity in individual neurons in a human brain, such studies can be done in monkeys. A simple example of binocular rivalry has been studied in a monkey by Nikos K. Logothetis and Jeffrey D. Schall, both then at M.I.T. They trained a macaque to keep its eyes still and to signal whether it is seeing upward or downward movement of a horizontal grating. To produce rivalry, upward movement is projected into one of the monkey's eyes and downward movement into the other, so that the two images overlap in the visual field. The monkey signals that it sees up and down movements alternatively, just as humans would. Even though the motion stimulus coming into the monkey's eyes is always the same, the monkey's percept changes every second or so.

Cortical area MT (which Semir Zeki calls in his chapter V5) is an area mainly concerned with movement. What do the neurons in MT do when the monkey's percept is sometimes up and sometimes down? (The researchers studied only the monkey's first response.) The simplified answer—the actual data are rather more messy—is that whereas the firing of some of the neurons correlates with the changes in the percept, for others the average firing rate is relatively unchanged and independent of which direction of movement the monkey is seeing at that moment. Thus, it is unlikely that the firing of

The Cheshire Cat Experiment

This simple experiment with a mirror illustrates one aspect of visual awareness. It relies on a phenomenon called binocular rivalry, which occurs when each eye has a different input from the same part of the visual field. Motion in the field of one eye can cause either the entire image or parts of the image to be erased. The movement captures the brain's attention.

To observe the effect, a viewer divides the field of vision with a mirror placed between the eyes (*a*). One eye sees the cat; the other eye a reflection in the mirror of a white wall or background. The viewer then waves the hand that corresponds to the eye looking at the mirror so that the hand passes through the area in which the image of the cat appears in the other eye (*b*). The result is that the cat may disappear. Or if the viewer was attentive to a specific feature before the hand was waved, those parts—the eyes or even a mocking smile—may remain (*c*).

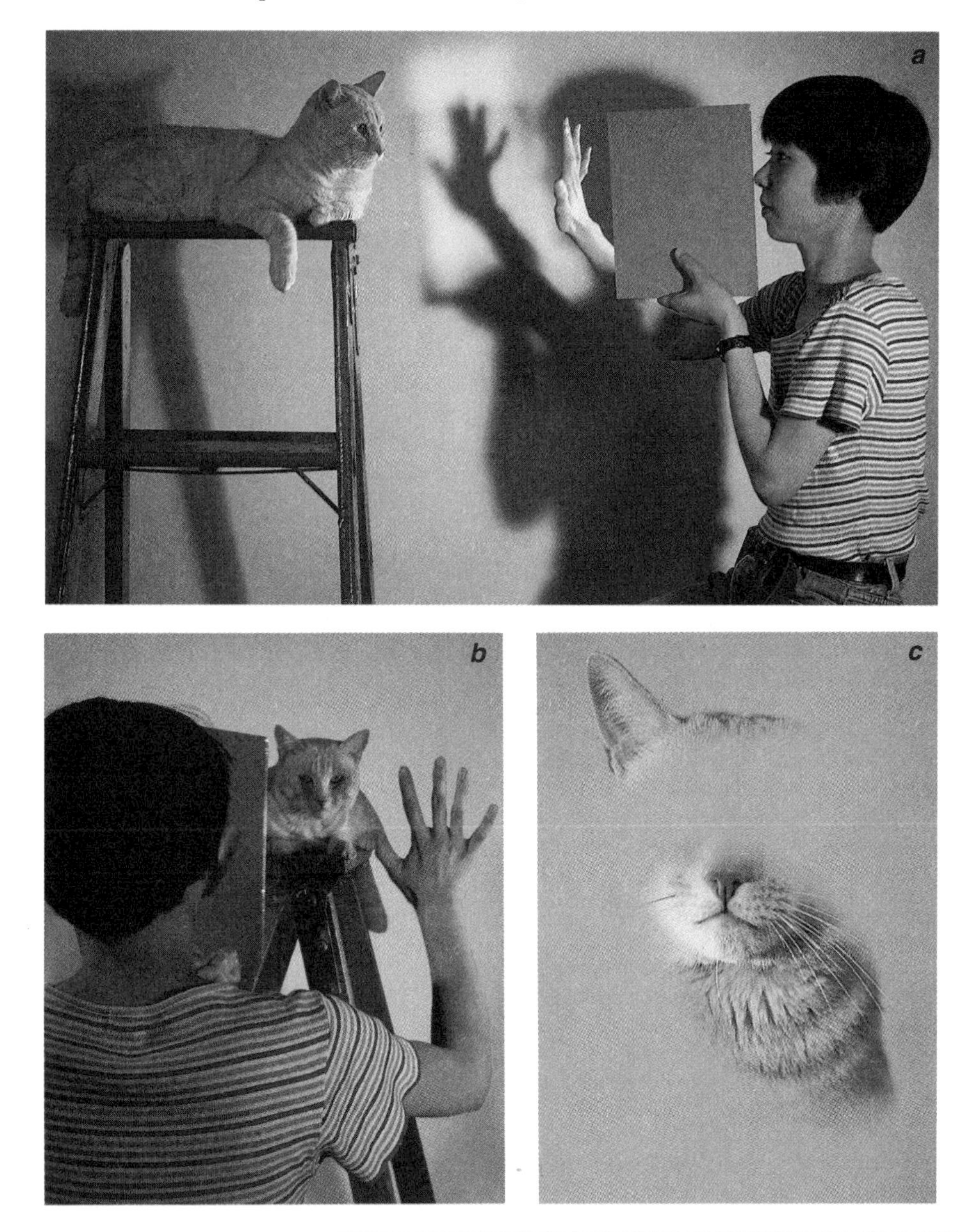

all the neurons in the visual neocortex at one particular moment corresponds to the monkey's visual awareness. Exactly which neurons do correspond remains to be discovered.

We have postulated that when we clearly see something, there must be neurons actively firing that stand for what we see. This might be called the activity principle. Here, too, there is some experimental evidence. One example is the firing of neurons in cortical area V2 in response to illusory contours, as described by Zeki. Another and perhaps more striking case is the filling in of the blind spot (see Figure 11.3). The blind spot in each eye is caused by the lack of photoreceptors in the area of the retina where the optic nerve leaves the retina and projects to the brain. Its location is about 15 degrees from the fovea (the visual center of the eye). Yet if you close one eye, you do not see a hole in your visual field.

Philosopher Daniel C. Dennett of Tufts University is unusual among philosophers in that he is interested both in psychology and in the brain. This interest is much to be welcomed. In a recent book, *Consciousness Explained,* he has argued that it is wrong to talk about filling in. He concludes, correctly, that "an absence of information is not the same as information about an absence." From this general principle he argues that the brain does not fill in the blind spot but rather ignores it.

Dennetts' argument by itself, however, does not establish that filling in does not occur; it only suggests that it might not. Dennett also states that "your brain has no machinery for [filling in] at this location." This statement is incorrect. The primary visual cortex (V1) lacks a direct input from one eye, but normal "machinery" is there to deal with the input from the other eye. Ricardo Gattass and his colleagues at the Federal University of Rio de Janeiro have shown that in the macaque some of the neurons in the blind-spot area of V1 do respond to input from both eyes, probably assisted by inputs from other parts of the cortex. Moreover, in the case of simple filling in, some of the neurons in that region respond as if they were actively filling in.

Thus, Dennett's claim about blind spots is incorrect. In addition, psychological experiments by Vilayanur S. Ramachandran [see "Blind Spots," SCIENTIFIC AMERICAN, May 1992] have shown that what is filled in can be quite complex depending on the overall context of the visual scene. How, he argues, can your brain be ignoring something that is in fact commanding attention?

Filling in, therefore, is not to be dismissed as nonexistent or unusual. It probably represents a basic interpolation process that can occur at many levels in the neocortex. It is, incidentally, a good example of what is meant by a constructive process.

How can we discover the neurons whose firing symbolizes a particular percept? William T. Newsome and his colleagues at Stanford University have done a series of brilliant experiments on neurons in cortical area MT of the macaque's brain. By studying a neuron in area MT, we may discover that it responds best to very specific visual features having to do with motion. A neuron, for instance, might fire strongly in response to the movement of a bar in a particular place in the visual field, but only when the bar is oriented at a certain angle, moving in one of the two directions perpendicular to its length within a certain range of speed.

It is technically difficult to excite just a single neuron, but it is known that neurons that respond to roughly the same position, orientation and direction of movement of a bar tend to be located near one another in the cortical sheet. The experimenters taught the monkey a simple task in movement discrimination using a mixture of dots, some moving randomly, the rest all in one direction. They showed that electrical stimulation of a small region in the right place in cortical area MT would bias the monkey's motion discrimination, almost always in the expected direction.

Thus, the stimulation of these neurons can influence the monkey's behavior and probably its visual percept. Such experiments do not, however, show decisively that the firing of such neurons is the exact neural correlate of the percept. The correlate could be only a subset of the neurons being activated. Or perhaps the real correlate is the firing of neurons in another part of the visual hierarchy that are strongly influenced by the neurons activated in area MT.

These same reservations apply also to cases of binocular rivalry. Clearly, the problem of finding the neurons whose firing symbolizes a particular percept is not going to be easy. It will take many careful experiments to track them down even for one kind of percept.

It seems obvious that the purpose of vivid visual awareness is to feed into the cortical areas concerned with the implications of what we see; from there the information shuttles on the one hand to the hippocampal system, to be encoded (temporarily) into long-term episodic memory, and on the

Figure 11.3 OPTICAL ILLUSION devised by Vilayanur S. Ramachandran illustrates the brain's ability to fill in missing visual information because it falls on the blind spot of the eye. When you look at the patterns of broken green bars, the visual system produces two illusory contours defining a vertical strip. Now shut your right eye and focus on the white square in the green series of bars. Move the page toward your eye until the blue dot disappears. Most observers report seeing the vertical strip completed across the blind spot, not the broken line. Try the same experiment with the series of just three red bars. The illusory vertical contours are less well defined, and the visual system tends to fill in the horizontal bar across the blind spot.

other to the planning levels of the motor system. But is it possible to go from a visual input to a behavioral output without any relevant visual awareness?

That such a process can happen is demonstrated by the remarkable class of patients with "blindsight." These patients, all of whom have suffered damage to their visual cortex, can point with fair accuracy at visual targets or track them with their eyes while vigorously denying seeing anything. In fact, these patients are as surprised as their doctors by their abilities. The amount of information that "gets through," however, is limited: blindsight patients have some ability to respond to wavelength, orientation and motion, yet they cannot distinguish a triangle from a square.

It is naturally of great interest to know which neural pathways are being used in these patients. Investigators originally suspected that the pathway ran through the superior colliculus. Recent experiments suggest that a direct albeit weak connection may be involved between the lateral geniculate nucleus and other cortical areas, such as V4. It is unclear whether an intact V1 region is essential for immediate visual awareness. Conceivably the visual signal in blindsight is so weak that the neural activity cannot produce awareness, although it remains strong enough to get through to the motor system.

Normal-seeing people regularly respond to visual signals without being fully aware of them. In automatic actions, such as swimming or driving a car, complex but stereotypical actions occur with little, if any, associated visual awareness. In other cases, the information conveyed is either very limited or very attenuated. Thus, while we can function without visual awareness, our behavior without it is rather restricted.

Clearly, it takes a certain amount of time to experience a conscious percept. It is difficult to determine just how much time is needed for an episode of visual awareness, but one aspect of the problem that can be demonstrated experimentally is that signals received close together in time are treated by the brain as simultaneous.

A disk of red light is flashed for, say, 20 milliseconds, followed immediately by a 20-millisecond flash of green light in the same place. The subject reports that he did not see a red light followed by a green light. Instead he saw a yellow light, just as he would have if the red and the green light had been flashed simultaneously. Yet the subject could not have experienced yellow until after the information from the green flash had been processed and integrated with the preceding red one (see Figure 11.4).

Experiments of this type led psychologist Robert Efron, now at the University of California at Davis, to conclude that the processing period for perception is about 60 to 70 milliseconds. Similar periods are found in experiments with tones in the auditory system. It is always possible, however, that the processing times may be different in higher parts of the visual hierarchy and in other parts of the brain. Processing is also more rapid in trained, compared with naive, observers.

Because it appears to be involved in some forms of visual awareness, it would help if we could discover the neural basis of attention. Eye movement is a form of attention, since the area of the visual field in which we see with high resolution is remarkably small, roughly the area of the thumbnail at arm's length. Thus, we move our eyes to gaze directly at an object in order to see it more clearly. Our eyes usually move three or four times a second. Psychologists have shown, however, that there appears to be a faster form of attention that moves around, in some sense, when our eyes are stationary.

The exact psychological nature of this faster attentional mechanism is at present controversial. Several neuroscientists, however, including Robert Desimone and his colleagues at the National Institute of Mental Health, have shown that the rate of firing of certain neurons in the macaque's visual system depends on what the monkey is attending to in the visual field. Thus, attention is not solely a psychological concept; it also has neural correlates that can be observed. Several researchers have found that the pulvinar, a region of the thalamus, appears to be involved in visual attention. We would like to believe that the thalamus deserves to be called "the organ of attention," but this status has yet to be established.

The major problem is to find what activity in the brain corresponds directly to visual awareness. It has been speculated that each cortical area produces awareness of only those visual features that are "columnar," or arranged in the stack or column of neurons perpendicular to the cortical surface. Thus, area V1 could code for orientation and area MT for motion. So far, as Zeki has explained, experimentalists have not found one particular region in the brain where all the information needed for visual awareness appears to come together. Dennett has dubbed such a hypothetical place "The Carte-

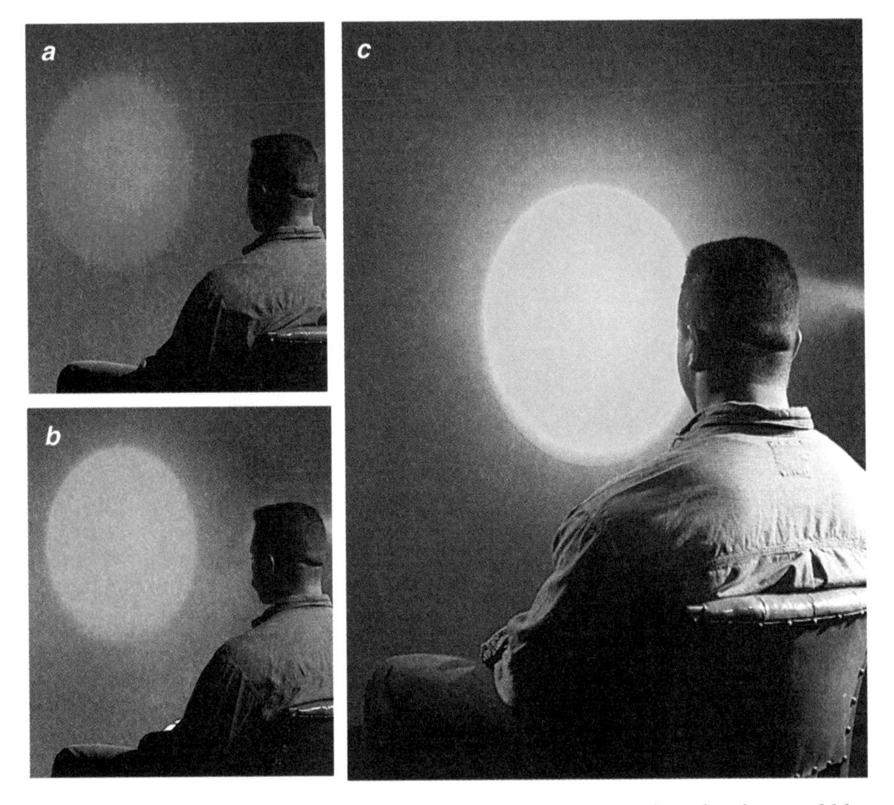

Figure 11.4 BRIEF FLASHES of colored light enable researchers to infer the minimum time required for visual awareness. A disk of red light is projected for 20 milliseconds (*a*), followed immediately by a 20-millisecond flash of green light (*b*). But the observer reports seeing a single flash of yellow (*c*), the color that would be apparent if red and green were projected simultaneously. The subject does not become aware of red followed by green until the length of the flashes is extended to 60 to 70 milliseconds.

sian Theater." He argues on theoretical grounds that it does not exist.

Awareness seems to be distributed not just on a local scale, as in some of the neural nets described by Hinton, but more widely over the neocortex. Vivid visual awareness is unlikely to be distributed over every cortical area because some areas show no response to visual signals. Awareness might, for example, be associated with only those areas that connect back directly to V1 or alternatively with those areas that project into each other's layer 4. (The latter areas are always at the same level in the visual hierarchy.)

The key issue, then, is how the brain forms its global representations from visual signals. If attention is indeed crucial for visual awareness, the brain could form representations by attending to just one object at a time, rapidly moving from one object to the next. For example, the neurons representing all the different aspects of the attended object could all fire together very rapidly for a short period, possibly in rapid bursts.

This fast, simultaneous firing might not only excite those neurons that symbolized the implications of that object but also temporarily strengthen the relevant synapses so that this particular pattern of firing could be quickly recalled—a form of short-term memory. (If only one representation needs to be held in short-term memory, as in remembering a single task, the neurons involved may continue to fire for a period, as described by Patricia S. Goldman-Rakic in Chapter 6, "Working Memory and the Mind.")

A problem arises if it is necessary to be aware of more than one object at exactly the same time. If all the attributes of two or more objects were represented by neurons firing rapidly, their attributes might be confused. The color of one might become attached to the shape of another. This happens sometimes in very brief presentations.

Some time ago Christoph von der Malsburg, now at the Ruhr-Universität Bochum, suggested that this difficulty would be circumvented if the neurons associated with any one object all fired in synchrony (that is, if their times of firing were correlated) but out of synchrony with those representing other objects. More recently, two groups in Germany reported that there does appear to be correlated firing between neurons in the visual cortex of the cat, often in a rhythmic manner, with a frequency in the 35- to 75-hertz range, sometimes called 40-hertz, or γ, oscillation.

Von der Malsburg's proposal prompted us to suggest that this rhythmic and synchronized firing might be the neural correlate of awareness and that it might serve to bind together activity in different cortical areas concerning the same object. The matter is still undecided, but at present the fragmentary experimental evidence does rather little to support such an idea. Another possibility is that the 40-hertz oscillations may help distinguish figure from ground [see "The Legacy of Gestalt Psychology," by Irvin Rock and Stephen Palmer; SCIENTIFIC AMERICAN, December 1990] or assist the mechanism of attention.

Are there some particular types of neurons, distributed over the visual neocortex, whose firing directly symbolizes the content of visual awareness? One very simplistic hypothesis is that the activities in the upper layers of the cortex are largely unconscious ones, whereas the activities in the lower layers (layers 5 and 6) mostly correlate with consciousness. We have wondered whether the pyramidal neurons in layer 5 of the neocortex, especially the larger ones, might play this latter role.

These are the only cortical neurons that project right out of the cortical system (that is, not to the neocortex, the thalamus or the claustrum). If visual awareness represents the results of neural computations in the cortex, one might expect that what the cortex sends elsewhere would symbolize those results. Moreover, the neurons in layer 5 show a rather unusual propensity to fire in bursts. The idea that the layer 5 neurons may directly symbolize visual awareness is attractive, but it still is too early to tell whether there is anything in it.

Visual awareness is clearly a difficult problem. More work is needed on the psychological and neural basis of both attention and very short term memory. Studying the neurons when a percept changes, even though the visual input is constant, should be a powerful experimental paradigm. We need to construct neurobiological theories of visual awareness and test them using a combination of molecular, neurobiological and clinical imaging studies.

We believe that once we have mastered the secret of this simple form of awareness, we may be close to understanding a central mystery of human life: how the physical events occurring in our brains while we think and act in the world relate to our subjective sensations—that is, how the brain relates to the mind.

Epilogue

Trouble in Mind

• • •

Jonathan Miller

Einstein once said that the most incomprehensible fact about nature was that it was comprehensible. It seems to me that the mysterious thing about nature is not that it is comprehensible but that it contains such a thing as comprehension at all, that is, the mind itself—the very idea!

Of course, ever since Darwin we have been able to consider the mind as something emerging from nature by virtue of certain arrangements developed through chance and necessity. Although Darwin is rightly credited with the discovery of the biological process that yielded such developments, their nature remained a mystery until the middle of this century. Thus, when early 19th-century physiologists brazenly invoked "material organization" as an explanation for both Life and Mind, it was still little more than hand-waving.

Today all that has changed. In the once unoccupied gap between the bare necessities of Matter and the enigmatic peculiarities of Mind, there is now an elaborate construction site of mediating concepts whose existence lends weight to the claim that the functions of mind are implemented by purely physical means. Neurons and their synapses *do* have the capacity to transact what is required to realize seeing and foreseeing. Conversely, at the other end of the psychophysical gap the image of mind itself has been radically redesigned. Under the influence of cognitive science, information theory and artificial intelligence, we can now envisage mental functions in a form that literally demands physical implementation.

The theoretical structures that reach toward one another across the great divide are developing so rapidly it is tempting to imagine that it's only a matter of time before they join up and that when that happens the terms once applied to mental life will be supplanted by a different sort of language altogether. In some quarters, this belief is already upheld as an article of faith; although paradoxically, among those who subscribe to it, the very concept of belief is regarded as questionable, along with many of the other psychological categories that figure in common parlance. Hence, the disparaging phrase "folk psychology." According to its critics, folk psychology is a hopelessly sloppy theory of human conduct, featuring terms such as "belief" and "desire," which are said to be not all that different from "phlogiston" and "caloric." Philosopher Paul M. Churchland insists that "we need therefore an entirely new kinematics and dynamics with which to understand human cognitive activity, one drawn perhaps from computational neuroscience and connectionist A.I. Folk Psychology could then be put aside in favor of descriptively more accurate . . . portrayals of the reality within."

No one can deny that the sciences to which Churchland refers will furnish more inclusive accounts than the ones we have at the moment. But that doesn't mean that folk psychology is nothing other than interim discipline. Nor is it a jury-rigged theory, good enough for ordinary folk and acceptable to experts as long as they are off duty. It is not and never was a theory, and to speak of it, as Churchland does, as an explanation "which fails on an epic scale" is to make a serious category mistake. The vernacular language of beliefs and desires is the expression of a form of life peculiar to creatures whose consciousness leads them to believe—yes, believe!—that others have experiences of the same kind.

In fact, it is hard to imagine how a radically different language would even begin to do a better job. It could and has been made more expressive, as any student of literature knows, and because of imaginative writers such as Shakespeare, say, or Proust, we can now convey psychological subtleties that were once inexpressible. A hard-core eliminativist would probably jeer that such revisions are doomed to failure and, like the epicycles added to Ptolemaic theory, are futile attempts to salvage a derelict system.

That, I venture to say, is nonsense. Consciousness may be implemented by neurobiological processes—how else?—but the language of neurobiology does not and cannot convey what it's *like* to be conscious. If, as philosopher Thomas Nagel says, there is something it's *like* to be a bat, there is something even more interesting it's like to be one of us, and the language of folk psychology, pidgin though it be in some respects, is the best medium for expressing this.

And that, of course, brings me to consciousness, which is quite rightly regarded as the most difficult problem in nature, more puzzling perhaps than the one which worried Einstein. And yet there are scientists prepared to insist that consciousness itself will yield to analysis, just as the problem of life has yielded. But there's the rub!—or rather the double bind. Although consciousness exists by virtue of some physical property of the brain, just as bioluminescence exists by virtue of some chemical property of certain specialized cells, it is *not*, as bioluminescence is, an observable property of living matter. It isn't a brain glow. Nor is it, on the other hand, an *invisible* property, less readily detectable than other biological processes. It is detectable to anyone who has it. The difficulty is that the method by which consciousness is detected is logically different from the method by which bioluminescence is detected. To put it bluntly, consciousness is not *detected* at all, because that would imply that it could pass *un*detected, and that doesn't make sense. *Your* consciousness may pass undetected by me, but *my* consciousness, if I have it at all, is self-evidently self-evident to me.

For that reason, we invoke a special kind of observation to which we can give the name "introspection." And yet introspection is not, as the word seems to imply, a peculiar form of gaze, directed at something inaccessible to the more typical forms of looking. On the contrary, introspection is one of the many forms consciousness itself can assume, so that it represents a significant part of what we are trying to explain. Indeed, the method by which we are acquainted with consciousness is so fundamentally different from the method by which we acquaint ourselves with brains that I suspect, as philosopher Colin McGinn does, that although we don't have to invoke anything other than brain—no magic that contravenes the laws of nature—we will never fully understand the connection.

This claim has been widely dismissed as frivolous obscurantism, foreclosing the possibility of further research and licensing the wilder forms of religious mysticism. It is nothing of the sort. There is obviously much more to be learned about the relationship between brains and minds, and it will be years, perhaps centuries, before we come up against the "cognitive closure" so courageously identified by Professor McGinn. The fact that such research is destined to describe an asymptotic curve, which approaches but never reaches the limit, does not preclude the necessity of our following it.

The Authors

GERALD D. FISCHBACH ("Mind and Brain") is Nathan Marsh Pusey professor of neurobiology and chairman of the department of neurobiology at Harvard Medical School and Massachusetts General Hospital. After graduating from Colgate University in 1960, he earned his medical degree at Cornell University Medical School in 1965 and received an honorary M.A. from Harvard University in 1978. Fischbach is also a member of the National Academy of Sciences, the National Institute of Medicine and the American Academy of Arts and Sciences. He is a past-president of the Society for Neuroscience and serves on several foundation boards and university advisory panels.

CARLA J. SHATZ ("The Developing Brain") is professor of neurobiology at the University of California, Berkeley, a position she took after many years at Stanford University. She graduated from Radcliffe College and received a master's degree in physiology from University College, London, and a Ph.D. in neurobiology from Harvard Medical School. Her studies of the development of connections in the mammalian visual system have gained her many honors, including her election to the American Academy of Arts and Sciences.

SEMIR ZEKI ("The Visual Image in Mind and Brain") is professor of neurobiology at the University of London. He obtained his doctorate from University College, London, and did his postdoctoral work at the National Institute of Mental Health in Washington, D.C., and at the University of Wisconsin at Madison. Zeki has also served as a visiting professor at several American and European universities. His particular interests center on the study of the anatomic and functional organization of the visual cortex in the monkey and in the human brain.

ERIC R. KANDEL and **ROBERT D. HAWKINS** ("The Biological Basis of Learning and Individuality") have collaborated on studies of the neurobiology of learning. Kandel is University Professor at the College of Physicians and Surgeons of Columbia University and senior investigator at the Howard Hughes Medical Institute. He received an A.B. from Harvard College, an M.D. from the New York University School of Medicine and psychiatric training at Harvard Medical School. Hawkins received a B.A. from Stanford University and a Ph.D. in experimental psychology from the University of California, San Diego. He is associate professor in the Center for Neurobiology and Behavior at Columbia.

ANTONIO R. DAMASIO and **HANNA DAMASIO** ("Brain and Language") have been investigating the neural basis of language and memory for the past two decades. Antonio Damasio is professor and head of the department of neurology at the University of Iowa College of Medicine and adjunct professor at the Salk Institute for Biological Studies in San Diego. He received his M.D. and doctorate from the University of Lisbon. Hanna Damasio also holds an M.D. from the University of Lisbon. She is professor of neurology and director of the Laboratory for Neuroimaging and Human Neuroanatomy at the University of Iowa.

PATRICIA S. GOLDMAN-RAKIC ("Working Memory and the Mind") has devoted her academic career to studying the neurobiology of memory and cognition. She received a Ph.D. from the University of California, Los Angeles, in 1963. Two years later she joined the Intramural Research Program of the National Institute of Mental Health. In 1979 she moved to the Yale University School of Medicine, where she is a professor of neuroscience. Goldman-Rakic sits on several national advisory boards and is a member of the National Academy of Sciences. She has also served as president of the Society for Neuroscience and her research focuses on identifying the neural mechanisms that carry out higher cortical functions in primates.

DOREEN KIMURA ("Sex Differences in the Brain") studies the neural and hormonal basis of human intellectual function. She is professor of psychology and honorary lecturer in the department of clinical neurological sciences at the University of Western Ontario in London. Kimura, a fellow of the Royal Society of Canada, received the 1992 John Dewan Award for outstanding research from the Ontario Mental Health Foundation and is the author of a book on neuromotor mechanisms in communication.

ELLIOT S. GERSHON and **RONALD O. RIEDER** ("Major Disorders of Mind and Brain") began collaborating on the study of mental illness at the Intramural Research Program of the National Institute of Mental Health. Gershon heads the program's Clinical Neurogenetics Branch and specializes in the population genetics and molecular genetics of normal and abnormal behavior. He graduated from Harvard Medical School in 1965, trained in psychiatry at the Massachusetts Mental Health Center and moved to the NIMH in 1969. Rieder graduated from Harvard Medical School in 1968 and trained in psychiatry at Albert Einstein College of Medicine. At the NIMH he conducted research on schizophrenia. He directs research and psychiatric residency training at Columbia University.

DENNIS J. SELKOE ("Aging Brain, Aging Mind"), who in 1988 received the Leadership and Excellence in Alzheimer's Disease Award from the National Institute on Aging, is co-director of the Center for Neurologic Diseases at Brigham and Women's Hospital in Boston. He is also professor of neurology and neuroscience at Harvard Medical School. Selkoe received his medical degree from the University of Virginia.

GEOFFREY E. HINTON ("How Neural Networks Learn from Experience") has worked on representation and learning in artificial neural networks for the past 20 years. In 1978 he received his Ph.D. in artificial intelligence from the University of Edinburgh. He is the Noranda Fellow of the Canadian Institute for Advanced Research and professor of computer science and psychology at the University of Toronto.

FRANCIS CRICK and **CHRISTOF KOCH** ("The Problem of Consciousness") share an interest in the experimental study of consciousness. Crick is the codiscoverer, with James Watson, of the double helical structure of DNA. While at the Medical Research Council Laboratory of Molecular Biology in Cambridge, he worked on the genetic code and on developmental biology. Since 1976, he has been at the Salk Institute for Biological Studies in San Diego. His main interest lies in understanding the visual system of mammals. Koch was awarded his Ph.D. in biophysics by the University of Tübingen. After spending four years at the Massachusetts Institute of Technology, he joined the California Institute of Technology, where he is associate professor of computation and neural systems. He is studying how single brain cells process information and the neural basis of motion perception, visual attention and awareness. He also designs analog VLSI vision chips for intelligent systems.

JONATHAN MILLER ("Epilogue"), physician, writer and director, produced the 1983 public television series "States of Mind" and the 1992 series "Madness."

Bibliography

2. The Developing Brain

Rakic, P. 1977. Prenatal development of the visual system in the rhesus monkey. *Philosophical Transactions of the Royal Society of London*, Series B, 278 (April 26): 245–260.

Miller, Kenneth D., Joseph B. Keller and Michael P. Stryker. 1989. Ocular dominance column development: Analysis and simulation. *Science* 245 (August 11): 605–615.

Shatz, Carla J. 1990. Competitive interactions between retinal ganglion cells during prenatal development. *Journal of Neurobiology* 21 (January): 197–211.

———. 1990. Impulse activity and the patterning of connections during CNS development. *Neuron* 5 (December): 745–756.

Goodman, Corey S., and Thomas M. Jessell, eds. 1992. Development. Special issue of *Current Opinion in Neurobiology* 2 (February).

3. The Visual Image in Mind and Brain

Zeki, S. M. 1978. Functional specialisation in the visual cortex of the rhesus monkey. *Nature* 274 (August 3): 423–428.

Weiskrantz, L. 1986. *Blindsight: A case study and implications*. Oxford University Press (Clarendon Press).

Livingstone, Margaret, and David Hubel, 1988. Segregation of form, color, movement, and depth: Anatomy, physiology, and perception. *Science* 240 (May 6): 740–749.

Zeki, S., and S. Shipp. 1988. The functional logic of cortical connections. *Nature* 335 (September 22): 311–317.

Edelman, Gerald M. 1990. *The remembered present: A biological theory of consciousness*. Basic Books.

Zeki, S. 1992. *A vision of the brain*. Blackwell Scientific Publications.

4. The Biological Basis of Learning and Individuality

Milner, Brenda. 1966. Amnesia following operation on the temporal lobes. In *Amnesia: Clinical, psychological and medicolegal aspects*, eds., C. W. M. Whitty and O. L. Zangwill. Butterworths.

Hawkins, R. D., T. W. Abrams, T. J. Carew and E. R. Kandel. 1983. A cellular mechanism of classical conditioning in aplysia: Activity dependent amplification of presynaptic facilitation. *Science* 219 (January 28): 400–405.

Nicoll, R. A., J. A. Kauer and R. C. Malenka. 1988. The current excitement in long-term potentiation. *Neuron* 1 (April): 97–103.

Squire, Larry L. 1992. Memory and the hippocampus: A synthesis from findings with rats, monkeys, and humans. *Psychological Review* 99 (April): 195–231.

5. Brain and Language

Klima, Edward S., and Ursula Bellugi. 1979. *The signs of language*. Harvard University Press.

Chomsky, Noam. 1986. *Knowledge of language: Its nature, origin, and use*. Greenwood Press.

Damasio, Hanna, and Antonio R. Damasio. 1989. *Lesion analysis in neuropsychology*. Oxford University Press.

Damasio, A. R., H. Damasio, D. Tranel and J. P. Brandt. 1990. Neural regionalization of knowledge access: Preliminary evidence. In *Cold Spring Harbor symposia on quantitative biology*, vol. LV, *The brain*. Cold Spring Harbor Laboratory Press.

Damasio, A. R. 1992. Aphasia. *New England Journal of Medicine* 326 (February 20): 531–539.

Fromkin, Victoria, and Robert Rodman. 1992. *An introduction to language*. Harcourt Brace Jovanovich College Publications.

6. Working Memory and the Mind

Baddeley, Alan. 1986. *Working memory*. Oxford University Press.

Goldman-Rakic, P. S. 1987. Circuitry of primate prefrontal cortex and regulation of behavior by representational memory. In *Handbook of physiology*, section 1, vol. 5, ed., Fred Plum. American Physiological Society.

Funahashi, Shintaro, Charles J. Bruce and Patricia S. Goldman-Rakic. 1989. Mnemonic coding of visual space in the monkey's dorsolateral prefrontal cortex. *Journal of Neurophysiology* 61 (February): 331–349.

Goldman-Rakic, Patricia S. 1991. Prefrontal cortical dysfunction in schizophrenia: The relevance of working memory. In *Psychopathology and the brain*, eds., Bernard J. Carroll and James E. Barrett, Raven Press.

7. Sex Differences in the Brain

DeVries, G. J., J. P. C. DeBruin, H. B. M. Uylings and M. A. Corner, eds. 1984. Sex differences in the brain: The relation between structure and function. In *Progress in brain research*, vol. 61. Elsevier.

Reinisch, J. M., L. A. Rosenblum and S. A. Sanders, eds. 1987. *Masculinity/femininity*. Oxford University Press.

Becker, Jill B., S. Marc Breedlove and David Crews, eds. 1992. *Behavioral endocrinology*. The MIT Press/Bradford Books.

8. Major Disorders of Mind and Brain

Carlsson, M., and A. Carlsson. 1990. Interactions between glutamatergic and monoaminergic systems within the basal ganglia—implications for schizophrenia and Parkinson's disease. *Trends in Neurosciences* 13 (July): 272–276.

Gershon, E. S., M. Martinez, L. R. Goldin and P. V. Gejman. 1990. Genetic mapping of common diseases: The challenges of manic-depressive illness and schizophrenia. *Trends in Genetics* 6 (September): 282–287.

Goodwin, Frederick K., and Kay Redfield Jamison. 1990. *Manic-depressive illness*. Oxford University Press.

Cooper, Jack R., Floyd E. Bloom and Robert H. Roth. 1991. *The biochemical basis of neuropharmacology*. Oxford University Press.

Chrousos, George P., and Philip W. Gold. 1992. The concepts of stress and stress system disorders: Overview of physical and behavioral homeostasis. *Journal of the American Medical Association 267 (March 4): 1244–1252.*

9. Aging Brain, Aging Mind

Coleman, P. D., and D. G. Flood. 1987. Neuron numbers and dendritic extent in normal aging and Alzheimer's disease. *Neurobiology of Aging* 8 (November/December): 521–545.

Weindruch, Richard, and Roy L. Walford. 1988. *The retardation of aging and disease by dietary restriction*. Charles C. Thomas.

Finch, Caleb E., and David G. Morgan. 1990. RNA and protein metabolism in the aging brain. In *Annual review of neuroscience*, vol. 13, eds., W. M. Cowan et al. Annual Reviews Inc.

Finch, Caleb E. 1990. *Longevity, senescence, and the genome*. University of Chicago Press.

Selkoe, Dennis J. 1991. The molecular pathology of Alzheimer's disease. *Neuron* 6 (April): 487–498.

Swaab, D. F. 1991. Brain aging and Alzheimer's disease: "Wear and tear" vs. "use it or lose it." *Neurobiology of Aging* 12 (July/August): 317–324.

10. How Neural Networks Learn from Experience

Rumelhart, David E., Geoffrey E. Hinton and Ronald J. Williams. 1986. Learning representations by back-propagating errors. *Nature* 323 (October 9):533–536.

Hinton, Geoffrey E. 1989. Connectionist learning procedures. *Artificial Intelligence* 40 (September): 185–234.

Hertz, J., A. Krogh and R. G. Palmer. 1990. *Introduction to the theory of neural computation*. Addison-Wesley.

Churchland, Patricia S., and Terrence J. Sejnowski. 1992. *The computational brain*. The MIT Press/Bradford Books.

11. The Problem of Consciousness

Rock, Irvin. 1984. *Perception*. Scientific American Library.

Jackendoff, Ray. 1987. *Consciousness and the computational mind*. The MIT Press/Bradford Books.

1990. *Cold Spring Harbor symposia on quantitative biology*, vol. LV, *The brain*. Cold Spring Harbor Laboratory Press.

Crick, Francis, and Christof Koch. 1990. Towards a neurobiological theory of consciousness. *Seminars in the Neurosciences* 2:263–275.

Churchland, Patricia S., and Terrence J. Sejnowski. 1992. *The computational brain*. The MIT Press/Bradford Books.

Sources of the Photographs

Museum of Modern Art, Mrs. Simon Guggenheim Fund: Figure 1.1
Janet Robbins, Harvard Medical School: Figure 1.2
Apostolos P. Georgopoulos, Society for Neuroscience: Figure 1.4
Marcus E. Raichle, Washington University School of Medicine: Figure 1.5

Little, Brown and Company: Figure 2.1
Carla J. Schatz: Figures 2.3 (*bottom*) and 2.6 (*bottom*)

Robert Prochnow: Figure 3.1
Semir Zeki: Figure 3.2 (*top right*) and "The World Seen Through a Damaged Cortex" (*top left* and *top right*)
Erlsbaum Associates: "The World Seen through a Damaged Cortex" (*bottom left*)
D. F. Benson, Archives of Neurology: "The World Seen through a Damaged Cortex" (bottom right)

Dan Wagner: Figure 4.1
Cambridge University Press: "Representation of the Surface of the Body in the Cortex" (*top*)

FPG International: Figure 5.1
Department of Neurology, PET Facility and Image Analysis Facility, University of Iowa: Figure 5.3
Marc Skinner: "Components of a Concept" (left)

Eric Hartman/Magnum: Figure 6.1
Harriet Friedman and Patricia S. Goldman-Rakic: Figure 6.4

Grandma Moses Properties Co., N.Y.: Figure 7.1

Edward Adamson and John Timlin: Figure 8.1
Nancy C. Andreasen, University of Iowa: Figure 8.3
Monte S. Buchsbaum, University of California at Irvine: Figure 8.4

FPG International: Figure 9.1
Dorothy G. Flood and Paul D. Coleman, University of Rochester: Figure 9.3
Dennis J. Selkoe: Figure 9.4
Ralph Warren Landrum, University of Kentucky: Figure 9.5
Robert P. Friedland, Case Western Reserve University, and John Wiley and Sons, Inc.: Figure 9.6

Geoffrey E. Hinton: "How a Neural Network Represents Handwritten Digits"

Herscovici/Artist Rights Society: Figure 11.1
Bettmann Archive: Figure 11.2
Jason Goltz: Figures 11.4 and "The Cheshire Cat Experiment"

INDEX

Page numbers in *italics* indicate illustrations.